Romantic Coffee Time
Coffee Shop's Graphic and Space Design

漫食光
——咖啡馆平面与空间设计

(哥) 卡洛斯·加西亚 胡书灵 编 张晨 译

辽宁科学技术出版社

COFFEE SHOPS AND DESIGNERS
咖啡馆与设计师

Have you ever wondered where creators take their ideas from? Well, most of the time I'm a creator and I can tell you that I get ideas from several places or moments, situations that may only mean something special for me, and that may be even absurd or really common for others. There are places, however, that may be considered temples of creation for almost every type of creator, from writers, poets and painters to designers, artists, architects and inventors, and one of those places are coffee-shops. I'm not sure if it's the magical beverages that we find there, or that everyone you see going in or already sitting inside have almost the same mood and have a clear goal for that very precise instant of their lives: to disconnect, or that the owners or creators of the best coffee shops in the world share their love for what they're doing and put it in every square centimeter of their shops, or maybe is the mixture of all those ingredients that make coffee shops unique, ever-lasting and always cool places to visit.

As an interior designer and co-creator of some really special coffee shops like 9¾ BOOKSTORE + CAFÉ and PERGAMINO (Medellín, Colombia) and CAFE BITE (Saudi Arabia), among others, I think there are no better projects to work in than the ones where you can print a part of who you are. And the best part of it is that your clients and their customers notice when a place is created with love, passion and knowledge.

The story behind 9¾ BOOKSTORE CAFE, for example, started almost just like that. By accident or luck, one special lady gave us a call and told us her dream. At first it sounded a little crazy and risky, but maybe because of that, it grabbed us all and made us fall in love with the project: a bookstore dedicated to children, located inside a street market in the mountains near Medellín (Colombia), 20 minutes away from the city, and with a coffee shop inside so adults could go with their kids and share together a moment surrounded by books. I call the idea crazy and risky because in this digital era, where printed material is questioned every day and in every possible media, and computers, smartphones and tablets rule how kids learn and read, and spend their time, the math just didn't sound that good. And, when you mix that with Medellín, a city where people are not so into reading, well…. you got to give her credit for her braveness.

After two or three meetings with our client, we were just infected with their security and the picture was clear: this new project would change us as a company and it'd make a huge impact in our city. And we were right, from that point on everything went smooth and nice, the ideas just came to us easily and the project took us to a moment in our lives where we liked to hide in small places, where the best possible moment was to share with our

你是否想过创意人员从何处获得灵感？我基本上就是一个创意从业者，对我来说一些场所和时刻都会带来灵感。这些情景可能对其他人来说十分普通，甚至有些怪异，但对我来说却有着特殊的意义。然而，有一些地方几乎对每类创意人员来说都是灵感的源泉，不论你是作家、诗人、画家，还是设计师、艺术家、建筑师和发明家。咖啡馆就是这样一种场所。我并不确定是因为咖啡馆供应咖啡这种神奇的饮料，还是每个进出咖啡馆的人都在走进或身处咖啡馆的那一时刻怀着几乎相同的情绪和明确的目标：要脱离周围的环境；抑或世界上最好的咖啡馆的店主或设计师与世人分享了他们对正在从事的事业的爱，并将这份爱呈现在咖啡馆的每寸空间里；又或者咖啡馆的那些元素组合之后使咖啡馆变得那样独特、经典而且让人百去不厌。

作为 9¾ 咖啡和佩尔加米诺咖啡（哥伦比亚，麦德林市）、CAFE BITE 咖啡馆（沙特阿拉伯）几家极具风格的咖啡馆的室内设计师和联合设计者，我认为那些能够映射出设计师自身特点的项目无疑是最好的。对设计师来说，最精彩的部分在于你的客户和他们的顾客都会注意到，这是一个充满了爱、热情和知识的设计。

以 9¾ 咖啡馆为例，它背后的故事就是这样开始的。出于偶尔的机会，一位特别的女士打电话联系到我们，讲述了她的梦想。起初，她的想法听起来有点疯狂，也存在风险，但或许正因为如此，它深深吸引了我们，让我们爱上了这个项目：一个坐落在距离市区 20 分钟路程的（哥伦比亚）麦德林市附近的山村购物街上，献给孩子们的书店。店内的咖啡馆方便成人与孩子一同前往，共享一段与书为伴的美好时光。我认为这个想法疯狂而有风险是因为在这个数字化的时代，平面媒体在每天都经受着来自各方的质疑，电脑、智能电话和平板电脑操纵着孩子们的学习方式和娱乐方式。在这种情况下，9¾ 咖啡馆的概念真的让人捏一把汗。尤其是把这个概念与麦德林市这个地点联系在一起，当地的人们对阅读的热情并不高涨……所以说，接受这个项目需要很大勇气。

与委托人进行了两三次会面之后，我们确认了项目的风险性，目标也变得明确：这个项目将改变 PLASMA NODO 工作室的历

parents an intimate evening in front of a good book, when we enjoyed stories with castles, speaking animals, wild nature, spaceships, ghosts, bad gangs, cars and planes, queens and kings, monsters and adventures…, It let us print in every graphic, in every piece of furniture or lamp, in every poster or wall finish, the parts of our lives when we were children, kids, teenagers and now adults. And yes, this is a true client-designer story, with no make-up. It happened just like that, as 'simple and easy' as I described, because some of the great things we learned is that there is no impossible project, there are good clients outside that trust our judgment as professionals and put in our hands the project of their lives, and that our job and effort are worth it.

The design process forces you to search deep inside the images in your memory and to apply your knowledge and your past experiences to understand specially the mistakes you've made so you don't make them again. This is a magical but sometimes frustrating moment and it makes you ask yourself how good can you really get to be, how much are you repeating yourself and your 'winner design formulas', are you heading the right way for the project, when will that white paper or empty screen have that image that takes to the instant you just say: I got it!.

Designing is like reading a good book you knew nothing about before, and it absorbs you into its pages: you take it everywhere with you, you can't stop thinking about it, not even after its finished and it makes you wonder what could've been different, how would the result had been if you had chosen a different way from the beginning, or how could that last pages could have ended with a better conclusion. For some of us, that may happen forever, and some of us can't even visit the places we created because we can't enjoy being there without criticizing ourselves for something we did or not did (believe me, I know many guys like those!). At that point, you have at least two choices: you can hate your job and quit to become something else, or you can sit, relax, drink a good cup of coffee and understand that you have the best possible work there is.

Carlos García
Co-founder and partner of PLASMA NODO

史，也将在城市中产生巨大的影响。事实证明我们是对的，项目从始至终进行得十分顺利，创意不断涌现，设计人员在项目的过程中似乎回到了童年，那时我们都喜欢藏在犄角旮旯，那时最幸福的时刻是晚上和父母安安静静地共读一本好书，那时我们喜欢故事里出现城堡、会说话的动物、大自然、宇宙飞船、鬼魂、坏帮派、汽车和飞机、皇后与国王、怪兽和冒险……我们可以在每个平面、每件家具、每张海报和墙面上展现出自己从儿时到青少年再到成人的生命历程。是的，就是这样一个客户与设计师之间发生的故事，真实发生了。事实就像我描述的这样"简单容易"，因为我们学到了一些重要的事情：世上没有不可能完成的项目。许许多多优秀的客户信任我们作为专业人士的判断，将自己生命一般的项目交到我们手上，我们付出再多努力都值得。

设计过程迫使你搜寻记忆深处的图像，运用所学知识和以往经验理解曾经犯过的错误，以求避免犯下同样的错误。这是一个神奇的过程，有时也会让人沮丧不已。你会扪心自问，自己究竟能做得多好？多大程度上是在重复自己的思路和"常胜设计模式"？是否找到了适合项目的正确的设计方向？白纸或空屏上何时能出现那个让你满意地拍手称赞的设计？

做设计就像读一本陌生的好书，它会将你深深吸引：你走到哪都会带着它，你无论做什么都想着它，即便是读完之后还会常常回味，如果一开始做了不同的选择，结果会有什么不同；或者最后几页是否可以交待一个更好的结果。对于一些人来说，这个过程会不断重复，而一些人甚至没法踏进自己设计的空间，因为他/她一定会因为设计中做了和没做的事情进行自我批判（相信我，这样的人我认识很多）！

卡洛斯·加西亚
PLASMA NODO 工作室联合创始人、合伙人

CONTENTS 目录

S TREET COFFEE SHOP
街头咖啡馆

ABARROTES DELIRIO 016
MEXICO CITY, MEXICO
Street Culture and Gastronomical Lifestyle

阿巴罗特·德利瑞欧咖啡馆
墨西哥，墨西哥城
街头文化与生活态度

BARISTACJA CAFE & BAKERY 020
BIALYSTOK, POLAND
A Design that Harmonizes Tastes and Smells of the Place

Baristacja 咖啡烘焙坊
波兰，比亚韦斯托克
味觉元素与嗅觉元素

WILD LIME 024
SOUTHAMPTON, UK
The Attitude of New World Places

野青柠咖啡馆
英国，南安普顿
新大陆地域的精神态度

ELSIE 026
AUCKLAND, NEW ZEALAND
Classic Vintage with Hand-crafted Neon Sign

埃尔希咖啡馆
新西兰，奥克兰
手工霓虹打造复古品牌形象

SIMPLE 028
TUNIS, TUNISIA
'In a World Full of Complexity, Being Simple Becomes an Extreme Sophistication.'

简单咖啡馆
突尼斯，突尼斯市
"在一个充满复杂的世界里，简单显得极为复杂"

OKTOKKI THE CAFE 032
SEOUL, SOUTH KOREA
Fun in Symbol & Icons

OKTOKKI 咖啡馆
韩国，首尔
图标符号的新趣味

CAFE IBERICO 034
QUERÉTARO, MEXICO
Custom Monogram Intensifies the Communication of the Shop Essence

伊比利亚咖啡馆
墨西哥，克雷塔罗市
特别设计的字体

NUDE 038
MOSCOW, RUSSIA
Extreme Simplicity and a Harmonized Space

Nude 咖啡馆
俄罗斯，莫斯科
极度简约设计营造和谐空间

THE BARISTA BOX 042
MANILA, PHILIPPINES
A Personal Ritual in a Mobile Café

The Barista Box 咖啡馆
菲律宾，马尼拉
移动咖啡馆里的个人仪式！

MONJO COFFEE 044
KUALA LUMPUR, MALAYSIA
Taking Cues from Metro Wayfinding Design

莫祖咖啡
马来西亚，吉隆坡
来自于地铁导视的灵感

SWEET TREAT 046
COPENHAGEN, DENMARK
Symbolic Reference to a Space Concept

甜品咖啡馆
丹麦，哥本哈根
字符间的非典型空间

PLOMBIERES ICE-CREAM CAFE 048
MOSCOW, RUSSIA
European café of House Economist

普隆比埃冰淇淋咖啡馆
俄罗斯，莫斯科
"经济学家宅邸"里的欧式咖啡馆

THEMED COFFEE SHOP
主题咖啡馆

THE ASSEMBLY 052
SINGAPORE
Highly Saturation Creates Concept Space

The Assembly 咖啡馆
新加坡
高饱和配色创造概念化空间

TALK' BY TALKIN' HEADS 068
GUATEMALA CITY, GUATEMALA
Elegant New York Fashion

Talk'By Talkin'Heads 咖啡馆
危地马拉，危地马拉市
优雅的纽约时尚风

LE PASTEL 080
CALIFORNIA, USA
Pet Dog Inspiration

宠物乐园咖啡馆
美国，加利福尼亚
狗狗带来的灵感

KOULTOURA 056
JAKARTA, INDONESIA
'We Bought a Zoo'

Koultoura 咖啡馆
印度尼西亚，雅加达
"我家买了动物园"

BARBER & SHACK 072
ABU DHABI, THE UNITED ARAB EMIRATES
When Barber Shop Meets Coffee Shops

Barber & Shack 咖啡馆
阿拉伯联合酋长国，阿布扎比
当理发店遇见咖啡馆

THE PLAYROOM 060
JAKARTA, INDONESIA
Fun Illustration in a Play Venue

游戏室咖啡馆
印度尼西亚，雅加达
游戏空间里的趣玩插画

CHE CAFE 076
BOULDER, USA
Everyone Loves Llamas!

Che Cafe 咖啡馆
美国，博尔德
人人都爱羊驼！

9¾ 064
MEDELLIN, COLOMBIA
Shh! This Is Not THE PLATFORM

9¾ 咖啡
哥伦比亚，麦德林市
嘘，这不是站台

THE GRUMPY CYCLIST 078
KUALA LUMPUR, MALAYSIA
A Cyclist-friendly Thirdwave Café

The Grumpy Cyclist 咖啡馆
马来西亚，吉隆坡
骑行者的异度空间

CONTENTS 目录

A RT COFFEE SHOP
艺术咖啡馆

THE LOCAL MBASSY   084
SYDNEY, AUSTRALIA
A Reflection of the Australian 1920s

THE LOCAL MBASSY 咖啡馆
澳大利亚，悉尼
二十世纪二十年代澳式风情

VINTAGE AND COFEE FOR MUSIC   096
BARCELONO, SPAIN
A World full of New Sensations and Tastes in Fresh Colours

音乐复古咖啡馆
西班牙，巴塞罗那
鲜明配色下的新感官

KINO CINEMA & ART CAFE   104
BUDAPEST, HUNGARY
Design with Images

吉纳影院和艺术咖啡馆
匈牙利，布达佩斯
设计的影像

DISHOOM   088
LONDON, UK
Inspiration from Beautiful Old Irani Cafés

DISHOOM 咖啡馆
英国，伦敦
灵感源于美丽的老式伊朗咖啡馆

CAFE RENZO BY RENZO PIANO   098
OSLO, NORWAY
Visual Identity of Architectural Thinking

伦佐·皮亚诺的伦佐咖啡馆
挪威，奥斯陆
富有建筑设计感的视觉形象

GAWATT COFFEE SHOP   090
YEREVAN, ARMENIA
Graffiti in Steampunk Style

GAWATT 咖啡馆
亚美尼亚，埃里温
蒸汽朋克风格的涂鸦

JAZZ HOUSE CAFE   100
BAKU, AZERBAIJAN
Jazz in Sketches

Jazz House 咖啡馆
阿塞拜疆，巴库
速写中的慵懒

VILLANDRY   094
LONDON, UK
Inspirations from Traditional Brasserie Menu Frames

维朗德里
英国，伦敦
传统啤酒铺菜单给予的灵感

MUSICAL CAFE 'ROYAL'   102
NIZHNY NOVGOROD, RUSSIA
The Colours of the Royal: Black & Gold

"皇家"音乐咖啡馆
俄罗斯，下诺夫哥罗德
最华丽的色彩：黑与金

INDEPENDENT COFFEE SHOP
独立咖啡馆

PERSILLADE-EAST MELBOURNE 108
GUADALAJARA, MEXICO
Multi-functions of a Café

东墨尔本——Persillade 咖啡馆
墨西哥，瓜达拉哈拉
咖啡馆的多功能性

MAMAN NYC 110
NEW YORK CITY, USA
'I was inspired by the waves of the sea'

马曼咖啡馆
美国，纽约
"我在海浪中找到了灵感"

GRASSHOPPER 114
MELBOURNE, AUSTRALIA
A Café with Nature in It

蚱蜢咖啡
澳大利亚，墨尔本
咖啡馆里的自然世界

DRUZHBA CAFE 116
MOSCOW, RUSSIA
The Image of the Bicycle 'Tandem'

德鲁日巴咖啡
俄罗斯，莫斯科
"tandem" 的自行车形象

C.A.P. 118
CHANTHABURI, THAILAND
The Renovation of a Historic Café

C.A.P. 咖啡馆
泰国，尖竹汶
老咖啡馆的新生

THE CUP 120
SFAX, TUNISIA
Modern Vintage Coffee Style from Tunisia

杯子咖啡馆
突尼斯，斯法克斯市
来自突尼斯的复古咖啡风格

LA MASCOTTE 124
SFAX, TUNISIA
Abstract, Minimal Perspective to a Modern Experience

马斯科特咖啡馆
突尼斯，斯法克斯市
模糊化的抽象设计理念

MICIO CAFFE 126
TAIWAN, CHINA
The Geometric Patterns for a Cat

米球咖啡
中国，台湾
猫的几何图形

WOHNZIMMA 130
SALZBURG, AUSTRIA
Vintage Sealed in Rubber Stamps

WohnZimma 咖啡馆
奥地利，萨尔斯堡
橡皮图章象征的复古氛围

FOUNDATION COFFEE HERTFORD 132
HERTFORD, UK
Kraft Paper and Brand Minimalism

赫特福德基金咖啡
英国，赫特福德郡
牛皮纸色彩下的简约风格

UMAMI ZEN CAFE 134
BRINDISI, ITALY
Visual Clash of Eastern and Western Cultures

屋纳米禅意咖啡
意大利，布林迪西
东西文化的视觉碰撞

CONTENTS 目录

BRANDED COFFEE SHOP
品牌咖啡馆

BABETTA CAFE — 138
MOSCOW, RUSSIA
Vintage with Memorable Details

芭贝特咖啡馆
俄罗斯，莫斯科
乱入的复古时尚感

THE CORNER — 152
MELBOURNE, AUSTRALIA
A Project of Twenty-Year-Plus Life Expectancy

街角咖啡馆
澳大利亚，墨尔本
设计方案的保质期：20 年

ATOMIC — 166
AUCKLAND, NEW ZEALAND
Inspiration Drawn from the Esoteric Tradition of Alchemy

原子咖啡馆
新西兰，奥克兰
灵感来自神秘的炼金术

COFFEE HOUSE LONDON — 142
LONDON, UK
Golden Embossing for London Culture

伦敦咖啡馆
英国，伦敦
金色雕花打造伦敦品质

MR TOMS — 156
AUCKLAND, NEW ZEALAND
Simple Urban Lifestyle

Mr Toms 咖啡
新西兰，奥克兰
大隐于市的独居格调

COFFEE SMITH — 168
TAIWAN, CHINA
A Storytelling Branding

咖啡匠
中国，台湾
品牌设计的故事性

HECHO CON AMOR — 146
CHIHUAHUA, MEXICO
Nostalgia from Grandmother's Kitchens

Hecho Con Amor 咖啡馆
墨西哥，奇瓦瓦
如祖母厨房般的亲切怀旧风潮

PARTOUT-EVERYWHERE — 160
TEL AVIV, ISRAEL
French Boutique Coffee Shop with Geometric Richness

Partout-Everywhere 咖啡馆
以色列，特拉维夫
法式精品与抽象几何的有机结合

MAJOR SPROUT — 172
AUCKLAND, NEW ZEALAND
The Appeal of A Wonderful Design

Major Sprout 咖啡馆
新西兰，奥克兰
设计赋予的感染力

LAUGHING MAN COFFEE — 150
NEW YORK, USA
'All Be Happy'

LAUGHING MAN 咖啡
美国，纽约
"一切皆快乐"

MONOLOG CAFE — 162
JAKARTA, INDONESIA
'A Conversation with One Self'

独白咖啡
印度尼西亚，雅加达
"与自己进行对话"

BA.RO.CO. — 176
AMBERG, GERMANY
Smart and Flexible Logo

ba.ro.co. 咖啡馆
德国，安贝格
智能灵活的标识设计

CAFE AND FASTFOOD
咖啡与简餐店

COFFEE SUPREME   180
WELLINGTON, NEW ZEALAND
Design of Irreverent Character

至上咖啡
新西兰，惠灵顿
设计里的玩世不恭

LITTLE DELI CO   188
SYDNEY, AUSTRALIA
Warmth like the Wooden Tone

小小食品公司
澳大利亚，悉尼
如木质感般的温馨

THE COHEMIAN COFFEE LOUNGE   200
IRUN, SPAIN
The Clashing of Blue and Pink

波西米亚咖啡馆
西班牙，伊伦
蓝与粉的色彩冲击

THE PAVILION CAFE   184
LONDON, UK
Café in Greenwich Park

THE PAVILION CAFE 咖啡馆
英国，伦敦
格林威治公园里的咖啡馆

COFFEE & KITCHEN   190
GRAZ, AUSTRIA
A Colourful World in Black and White

咖啡与餐厅
奥地利，格拉茨
黑白背景下的色彩世界

MAHLWERK   202
LANGENFELD, GERMANY
Simple Design and Quality Life

MAHLWERK 咖啡食品店
德国，朗根费尔德
简单设计，精致生活

LA CASITA CAFE   194
TENERIFE, SPAIN
The Colours of Summer

小屋咖啡馆
西班牙，特纳利夫
夏日的颜色

BÖMARZO   206
ASTURIAS, SPAIN
'o' The Dieresis

博马尔佐咖啡馆
西班牙，阿斯图里亚斯
"0"形隔音符号

MILK & HONEY   198
CHATTANOOGA, USA
Illustrated Vitality

牛奶与蜂蜜咖啡馆
美国，查特怒加市
插画的活力

CAFE OF JULIA VYSOTSKAYA   208
MOSCOW, RUSSIA
Illustrations from Julia Vysotskaya's Books

朱莉娅·维斯托斯卡亚的咖啡馆
俄罗斯，莫斯科
朱莉娅·维斯托斯卡亚书中的插画

CONTENTS 目录

URBAN HEART — 210
KYIV, UKRAINE
The Simplicity of a Black and White Palette

城市心脏咖啡馆
乌克兰,基辅
黑白系的简约态度

LEGACY ROASTERS — 220
GUADALAJARA, MEXICO
Symbolic Use of Graphics

传奇风味咖啡馆
墨西哥,瓜达拉哈拉
图形设计的象征印记

PARC PANTRY — 232
WALES, UK
Rustic yet Stylish Aesthetic

帕克咖啡食品店
英国,威尔士
犀利的现代设计

TIME OFF — 212
ANTWERP, BELGIUM
Relaxing and Refreshing Mint Green

"休息时间"咖啡馆
比利时,安特卫普
薄荷绿的清新风格

COLOCHATE CAFETERIA — 224
SANTIAGO, CHILE
A Welcoming Sense of Belonging

Colochate 咖啡馆
智利,圣地亚哥
温暖的归属感

KAFFEE KANN ICH. — 234
HANNOVER, GERMANY
The Visual Appeal of a Simple Design

Kaffee kann ich. 咖啡馆
德国,汉诺威
简洁设计的视觉魅力

PALEM CAFE RESTAURANT — 214
JAKARTA, INDONESIA
Strong, Unique and Balanced

帕莱姆咖啡餐厅
印度尼西亚,雅加达
强大、独特、平衡

CAPULUS TABERNAM — 228
MONTERREY, MEXICO
Nordic Impression

Capulus Tabernam 咖啡馆
墨西哥,蒙特雷
北欧印象

SANDWICH FACTORY — 236
SYDNEY, AUSTRALIA
The Beauty of Simplicity and Nature

墙洞三明治工厂
澳大利亚,悉尼
朴素而自然的美

CAFE CROWBAR — 218
BRNO, CZECH REPUBLIC
A Strange Angle of Typeface Design

撬棍咖啡馆
捷克,布尔诺
字体设计的奇特角度

SMOKE & ROAST — 230
LONDON, UK
The Colours of Food

Smoke & Roast 咖啡馆
英国,伦敦
食物的色彩

COMO ST CAFE — 238
AUCKLAND, NEW ZEALAND
A Magical World of Colours

科摩街咖啡馆
新西兰,奥克兰
色彩的魔法空间

OTHER CREATIVE CAFE
其他创意咖啡馆

MOTO MOTO CAFE   240
STAVROPOL, RUSSIA
Cheerful Madness

Moto Moto 咖啡馆
俄罗斯，斯塔夫罗波尔
欢快的疯狂

CARDINAL CAFE   250
BERKSHIRE, UK
A Cardinal Bird for Inspiration

北美红雀咖啡馆
英国，伯克郡
设计灵感来自北美红雀

BRONUTS   262
WINNIPEG, CANADA
A Tongue in Cheek Personality

Donuts & Coffee 咖啡馆
加拿大，温尼伯
有趣的脸谱

MANTRA RAW VEGAN   242
MILAN, ITALY
Taking Away a Seed of Rebirth

曼特拉生素食主义
意大利，米兰
"在心中埋下一颗重生的种子"

MISEGRETA   252
GUADALAJARA, MEXICO
A 'Secret Mix'

MISEGRETA 咖啡馆
墨西哥，瓜达拉哈拉
"秘密的混合"

LE PASTEL   266
NEW YORK CITY, USA
Perfect Balance in a Combination of Colours

彩色粉笔咖啡馆
美国，纽约
色彩选择中的完美平衡

CASA E CAFFE MOKADOR   246
FAENZA, ITALY
The First Concept Bar of Mokador

Casa e caffe Mokador 咖啡馆
意大利，法恩扎
第一间 Mokador 概念咖啡馆

LOUIS CHARDEN   256
YEREVAN, THE REPUBLIC OF ARMENIA
Illustration Design with Intriguing Stories

Louis Charden 咖啡馆
亚美尼亚共和国，埃里温
有故事的插画设计

{t} TELEGRAPHE CAFE   270
NEW YORK, USA
Typography and Symbols

{ t } 电报咖啡馆
美国，纽约
字型与符号的发想

CLOUD.COFFEE SHOP   260
MILAN, ITALY
Design with Elements from Sky

云。咖啡馆
意大利，米兰
天空中的元素

CAFE M   272
SEOUL, KOREA
The Symbolic Meaning of the Letter M

M 咖啡
韩国，首尔
M 图案的象征意义

STREET COFFEE SHOP

街头咖啡馆

MEXICO CITY, MEXICO

ABARROTES DELIRIO

Street Culture and Gastronomical Lifestyle

阿巴罗特·德利瑞欧咖啡馆 / 墨西哥，墨西哥城

街头文化与生活态度

The objective was to communicate Abarrotes Delirio's philosophy, based on practicality, simplicity and on the organic provenance of their ingredients, through its identity and interior design, which in turn was developed by Habitación 116. The careful selection of the products on offer is depicted in the graphic style developed for the brand, where each element serves a specific purpose to complement this lifestyle. The contrast between the opposing concepts of a corner shop and a gourmet one is resolved through a neutral and pristine base that can cater for all forms and colours which are naturally present in the products, thus effectively communicating the concept of closeness, authenticity and selectiveness.

Design agency: SAVVY Studio Client: Abarrotes Delirio

本次设计的目的是通过品牌形象设计及室内设计，传达店铺务实、简约、崇尚有机食材的品牌精神。精心挑选的店内产品呈现在专门为店铺设计的平面方案中，每个元素的选用都以目标明确的形式，使店铺主张的这一生活态度更加丰满。中立的原始地形适应产品中自然呈现的各种形式和配色，使街角小店和美食店的对立设计概念之间的反差得以缓和，有效地传递出亲密、真诚和精挑细选的经营理念。

设计机构：Savvy 工作室 委托方：阿巴罗特·德利瑞欧咖啡馆

BIALYSTOK, POLAND

BARISTACJA CAFE & BAKERY

A Design that Harmonizes Tastes and Smells of the Place

Baristacja 咖啡烘焙坊 / 波兰,比亚韦斯托克

味觉元素与嗅觉元素

Baristacja is a place where you can drink great coffee and taste homemade bread. All of the beverages and meals are being prepared with lots of love and passion. The task was to create branding and interior design for the café which would harmonize with all of the tastes and smells of the place. It was also important to create interior which would change its atmosphere during the day - from bright, energetic breakfast place, through comfortable lunch / midday meeting space, finishing as evening calm and cozy café serving various beverages. It was possible by using different light scenarios - bright sunlight passing by big windows and warm artificial lighting. Used spruce wood refers to the naturalness of the place and food served. The place is located in center of Białystok, Poland, next to main street aggregating many morning commuters. It works as the 'pit-stop' to grab a coffee and some baked goods, thats' why the name is Baristacja - standing for barista + stacja (station in Polish).

Designer: Joanna Karwowska, Marek Ejsztet, Piotr Matuszek
Photography: Rafał Kłos Client: Essenza Catering

人们可以在 Baristacja 咖啡烘焙坊喝到优质咖啡，品尝自制面包。所有的饮品和餐食都倾注了巨大的爱心和热情。本项目的目标是为咖啡馆打造品牌形象和室内设计方案，整体呈现所有味觉和嗅觉元素。同时打造一个能够在日间改变室内气氛的空间设计——实现从轻快、充满活力的早餐到舒适的午餐/日间会面空间，以及供应多种不同饮品的沉静、惬意的夜间咖啡馆。这样的气氛转变是通过不同的灯光场景设计实现的——穿透大窗户的明亮光线以及温和的人造光。再利用云杉木反映咖啡馆和食品的天然和环保。店铺位于波兰比亚韦斯托克市中心，紧邻主街，清晨时分有大量通勤者在此聚集。这里是选购咖啡和烘焙食品的绝佳"起点"，店铺的名字也由此而来——"Baristacja"是"barista"（咖啡调配师）+"stacja"（波兰语中的车站）的合成词。

设计师：乔安娜·卡沃斯卡，马雷克·艾杰特，彼得·麦楚塞克 摄影：拉法尔·克罗斯 委托方：Essenza 餐饮公司

SOUTHAMPTON, UK
WILD LIME

The Attitude of New World Places

野青柠咖啡馆 / 英国，南安普顿

新大陆地域的精神态度

& SMITH and We All Need Words developed Wild Lime's brand, visual identity and tone of voice from scratch. The brand is inspired by the attitude of New World places like California, Cape Town and Sydney. The idea was to bring a little of the coffee and brunch culture that's come to London from places like Auckland and Melbourne to the high street.

Design agency: & SMITH Designer: Sam Kang
Photography: & SMITH

& SMITH 设计公司与 We All Need Words 工作室共同创造了野青柠咖啡馆的品牌形象、视觉形象和基调。该品牌起源于加州、开普敦和悉尼等新大陆地域的精神态度,将奥克兰和墨尔本等地传入伦敦的咖啡和早午餐文化传播到繁华商业街上。

设计机构:& SMITH 设计公司 设计师:山姆·康
摄影:& SMITH 设计公司

AUCKLAND, NEW ZEALAND

ELSIE

Classic Vintage with Hand-crafted Neon Sign

埃尔希咖啡馆 / 新西兰，奥克兰

手工霓虹打造复古品牌形象

Tasked with brightening a small space on a grey street, Fuman created a retro-inspired brand through the use of a simple, classic neon sign of hand-crafted type. Elsie's equally uncomplicated interior showcases accents of canary yellow, radiating both light and warmth with a familiar charm.

Design agency: Fuman Design Studio Designer: Jon Chapman-Smith Photography: Jon Chapman-Smith Client: Elsie

本项目中，孚曼设计工作室受邀为灰色调街道上的一处小空间注入光明，利用简洁、经典的手工霓虹标志打造出一个以复古为灵感的品牌形象。埃尔希咖啡馆的室内展示延续了简约的设计风格，以淡黄色为主，展现轻快、温馨的魅力。

设计机构：孚曼设计工作室 设计师：乔恩·查普曼·史密斯 摄影：乔恩·查普曼·史密斯 委托方：埃尔希咖啡馆

TUNIS, TUNISIA

SIMPLE

*'In a World Full of Complexity,
Being Simple Becomes an Extreme Sophistication.'*

简单咖啡馆 / 突尼斯，突尼斯市

"在一个充满复杂的世界里，简单显得极为复杂"

"In a world full of complexity, being simple becomes an extreme sophistication." Ramdam Agency built the design of a new food & coffee shop based on this philosophy : A sleek and minimalistic design starting from the brand itself to the store design and the digital material.

Design agency: Ramdam Agency Creative director: Thierry Silvestri
Designer: Thierry Silvestri, Kaïs Belaïba, Selim Ben Hadj Yahia
Photography: Kaïs Belaïba Client: Simple Food Shop

"在一个充满复杂的世界里，简单显得极为复杂。"拉姆丹设计公司以此为基础打造了全新的食品和咖啡馆。时尚而简约的设计方案形成于品牌自身，延续到店面设计和数码元素中。

设计机构：拉姆丹设计公司 创意总监：蒂埃里·西尔维斯特里 设计师：蒂埃里·西尔维斯特里，卡伊斯·贝来巴，塞利姆·本·海至·雅希亚 摄影：卡伊斯·贝来巴 委托方：简单食品店

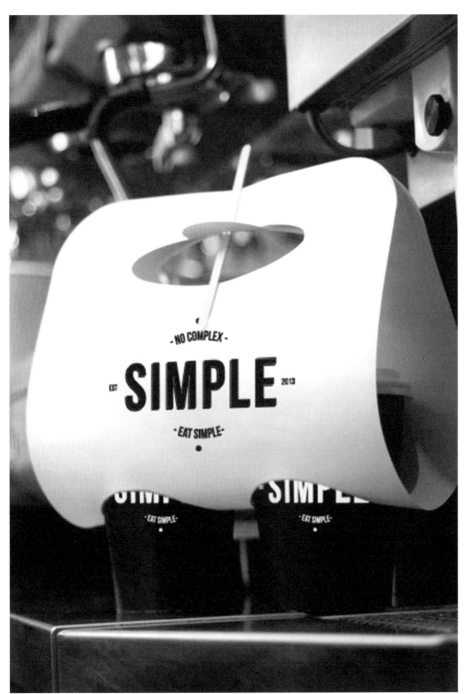

OKTOKKI® THE CAFÉ

 RICE 米

 SSAAM

 BAAN

KOREAN FINGER FOOD

 SUPER SALAD

 FRESH COFFEE

 REFRESHING TEA

 ECO JUICE

 NATURAL ICE-FLAKES

In order to show OKTOKKI's powerful point that they use own fusion recipes for traditional foods to access trendy customers, ANZI & ONASUP used very simple symbol & icons within many metaphors such as grill, rice, Ssam & Ban and so on. In additional, they designed packaging with only black & white colour to keep cool image of the shop.

Design agency: ANZI & ONASUP Photography: ANZI & ONASUP Client: OKTOKKI The Cafe

OKTOKKI 使用独创配方制作传统食物，满足现代顾客的饮食需求。为了展现这一强大之处，设计师选择非常简约的，代表烧烤、大米和辣豆酱等元素的符号和图标。此外，设计师还完成了黑白色系的包装设计，延续时尚的整体风格。

设计机构：ANZI & ONASUP 工作室 摄影：ANZI & ONASUP 工作室 委托方：OKTOKKI 咖啡馆

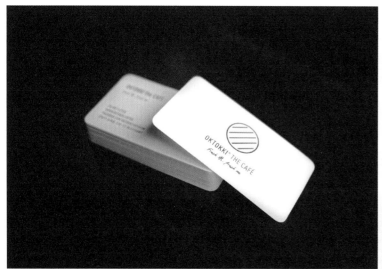

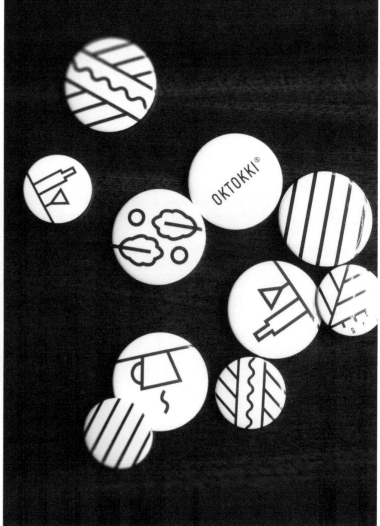

SEOUL, SOUTH KOREA

OKTOKKI THE CAFE

Fun in Symbol & Icons

OKTOKKI 咖啡馆 / 韩国, 首尔

图标符号的新趣味

QUERÉTARO, MEXICO

CAFÉ IBERICO

Custom Monogram Intensifies the Communication of the Shop Essence

伊比利亚咖啡馆 / 墨西哥，克雷塔罗市

特别设计的字体

Brand design project for 'Café Ibérico', coffee shop located in the state of Guanajuato Mexico, dedicated in bakery and the preparation of cold and hot beverages. The use of a simple typography and a monogram specially designed, intensifies the communication of the essence of the place: taste and comfort.

Design agency: ESTUDIO INSURGENTE (Mexico) Designer: Igor Orozco / Gilberto Castillo / Rodrigo González Photography: Igor Orozco Client: CAFÉ IBÉRICO

本案是为坐落在墨西哥瓜纳华托州的"伊比利亚咖啡馆"进行的品牌设计项目。咖啡馆供应烘焙点心和冷热饮品。简洁的版式设计和特别设计的字体强化了店内空间的精髓传递：品味和舒适度。

设计机构：INSURGENTE（墨西哥）工作室 设计师：伊戈尔·奥罗斯科 / 吉尔伯托·卡斯蒂略 / 罗德里戈·冈萨雷斯 摄影：伊戈尔·奥罗斯科 委托方：伊比利亚咖啡馆

MOSCOW, RUSSIA

NUDE

Extreme Simplicity and a Harmonized Space

Nude 咖啡馆 / 俄罗斯，莫斯科

极度简约设计营造和谐空间

The interior design was informed by the materiality of the existing space. The concept of identity keeps the same idea — simplicity and no excess. All the elements have transparent core, not hiding the interior surroundings but filling it up as a single whole.

Designer: Denis Sharypin Photography: Asya Baranova Interior design: Asya Baranova Client: Artur and Zara Bersirov

本案的室内设计根据空间现有的材料展开,品牌概念延续相同的理念——极度简约。所有元素都呈现透明的特质,使整个室内空间通透和谐。

设计机构:丹尼斯·沙里皮工作室 摄影:阿斯亚·巴拉诺娃 室内设计师:阿斯亚·巴拉诺娃 委托方:阿图尔·博斯洛夫,莎拉·博斯洛夫

MANILA, THE PHILLIPPINES

THE BARISTA BOX

A Personal Ritual in a Mobile Café

The Barista Box 咖啡馆 / 菲律宾，马尼拉

移动咖啡馆里的个人仪式

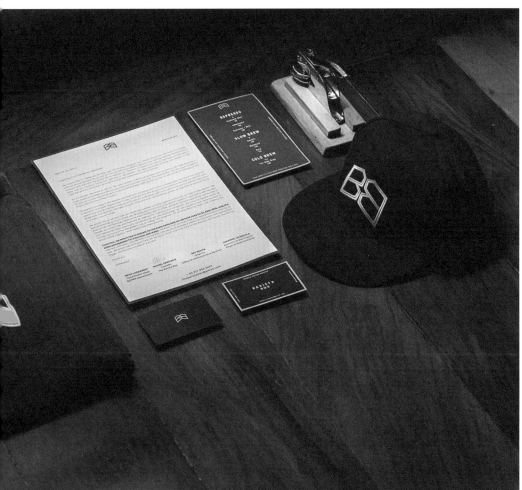

Barista Box is the Phillippines' first mobile coffee concept. It's old fashioned and the owners fuss about their coffee, but only because they believe that each drinking experience is a personal ritual. They focus on the classics in order to help people rediscover why life is too short for bad coffee.

Designer: Isai Araneta Client: The Barista Box Photography: Miguel Santiago, Kimi Juan

The Barista Box 是菲律宾第一家移动咖啡馆。这里提供老式咖啡，店家对产品品质极为重视，深信每杯咖啡的体验都是一场关于个人的仪式。店铺的经典咖啡理念帮助人们重新发现优质咖啡对生活的重要性。

设计：伊赛·阿拉内塔 委托方：The Barista Box 咖啡馆 摄影：米格尔·圣地亚哥，基米·胡安

KUALA LUMPUR, MALAYSIA

MONJO COFFEE

Taking Cues from Metro Wayfinding Design

莫祖咖啡 / 马来西亚，吉隆坡

来自于地铁导视的灵感

Rice Creative were asked to develop a concept, brand, and visual identity for a new coffee franchise, which would originate in Kuala Lumpur, Malaysia. This new franchise should offer a competitive option from some of the international coffee shop chains in the city. Rice Creative conducted research about Kuala Lumpur, drank a lot of coffee, and discovered a unique coffee culture in KL. They found an opportunity to create a coffee shop which offered a little more and felt like one of a kind, while being relatively simple to replicate. This coffee shop should offer quick convenience to the busy citizens of the metropolis. It should also offer a place to relax and take a break or work away for hours. Taking cues from metro wayfinding design, Rice Creative developed icons for nearly everything in the entire location, from coffee cups, to sandwiches, to real live coffee trees. They made icons of the exact espresso machine Monjo uses as well as the grinder, and blender. Even the light fixtures icons are signature to Monjo. These icons will be used in Monjo's advertising, giving the public a taste of all the things they can expect to find at Monjo.

Design agency: Rice Creative Creative director: Joshua Breidenbach & Chi-An De Leo Designer: Nguyen Huynh & Truc Dinh Photography: Joshua Breidenbach Client: Monjo Coffee

设计师受邀为一家新咖啡馆进行概念、品牌和视觉形象设计。该品牌源自马来西亚的吉隆坡,这家新店提供与城内的国际咖啡连锁店旗鼓相当的产品与服务。经过对吉隆坡本地资源的深入研究,以及对咖啡产品的评估,设计师发现了吉隆坡特有的咖啡文化,寻找到一个与众不同又易于复制的咖啡馆模式。在这个模式中,咖啡馆为繁忙的城市居民提供便捷的服务,同时营造一个让顾客休闲放松或者潜心工作的氛围。设计师从地铁导视设计中汲取灵感,创造了包括咖啡杯、三明治和咖啡树在内的绝大部分店内标识。设计师按照店内实际使用的浓缩咖啡机、研磨机和搅拌机制作了准确的标识。即便是灯具标识都极具店铺特点。这些标识都被使用在莫祖咖啡的广告中,以微观的视角向人们呈现出店内的真实情况。

设计机构:莱斯创意工作室 创意总监:约书亚·布赖登巴赫,吉安·德里奥设计师:黄阮(音译),特吕克·丁摄影:约书亚·布赖登巴赫 委托方:莫祖咖啡

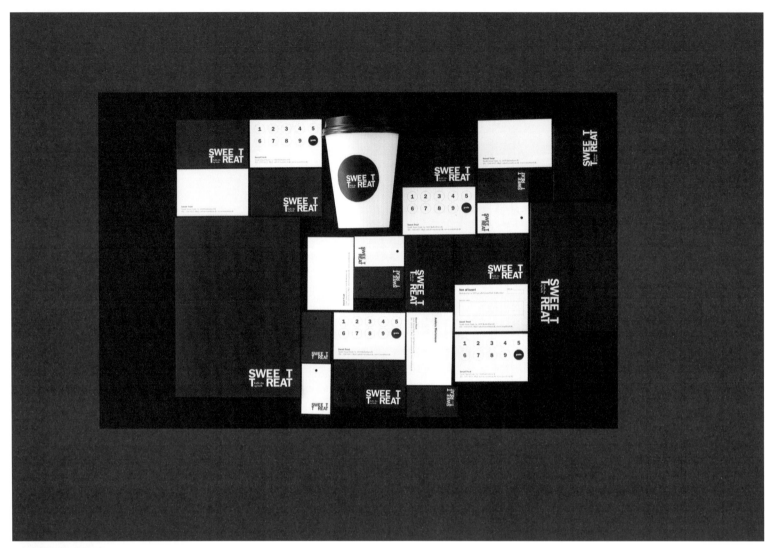

The simple and aestehtic logotype exhibit both quality and courtesy. The atypical space between the charaters are carefulley set as a symbolic reference to the concept of the break: The break you take when you visit the café, and the pleasure break you experience when tasting something of superlative quality.

Design agency: Re-public Designer: Stina Nordquist Photography: Jenny Nordquist & Christoffer Hald Client: Sweet Treat

简洁、美观的标识设计展现出质感和谦恭的姿态。字符间的非典型空间体现精心设计,象征性呼应休息时光的概念:在咖啡馆停留的休息时间以及品尝优质食材时获得的愉悦体验。

设计机构: Re-public 工作室 设计师: 斯蒂娜·诺德奎斯特 摄影: 詹妮·诺德奎斯特,克里斯托弗·哈尔德 委托方: 甜品咖啡馆

COPENHAGEN, DENMARK

SWEET TREAT

Symbolic Reference to a Space Concept

甜品咖啡馆 / 丹麦，哥本哈根

字符间的非典型空间

MOSCOW, RUSSIA

PLOMBIERES ICE-CREAM CAFE

European Café of House Economist

普隆比埃冰淇淋咖啡馆 / 俄罗斯，莫斯科

"经济学家宅邸"里的欧式咖啡馆

Design of the logo, corporate identity and website for Scenario café in Moscow. Scenario is a European cafe that opened on the first floor of a magnificent 'House Economist' on Tverskaya Street in Moscow. Here one can find white walls, colourful couches and chairs, bright illustrations of Benoit Fizeau and softly rock ballads, which create the necessary relaxing atmosphere. The name of the cafe has nothing to do with the cinema — we are talking more about life scenarios: a cup of coffee and a chat with friends in a pleasant atmosphere, delicious and inexpensive lunch with colleagues.

Design agency: Province Design Studio Designer: Elena Trofimova, Pavel Bogdanov, Evgeny Pakhomov Photography: Alexaner Yarysh Client EF Restaurants LLC

本项目涉及莫斯科"场景"咖啡馆的品牌标识、企业形象及网站设计。"场景"咖啡馆是一家欧式咖啡馆，位于莫斯科特维尔大街上宏伟的"经济学家宅邸"一楼。这里有白墙，缤纷的沙发和椅子，明快的插画以及轻摇滚民谣，氛围轻松惬意。咖啡馆的名字与电影中的场景概念并无关系，它的灵感更多地来自于生活场景：就着一杯咖啡与友人愉快地交谈，或者与同事共享经济美味的午餐。

设计机构：Province 设计工作室 设计师：叶夫根尼·帕霍莫夫 摄影：亚历山大·亚雷什 委托方：EF 餐厅责任有限公司

THEMED COFFEE SHOP

主题咖啡馆

SINGAPORE

THE ASSEMBLY

Highly Saturation Creates Concept Space

The Assembly 咖啡馆 / 新加坡

高饱和配色创造概念化空间

The Assembly is a multi-label lifestyle store targeted at men's fashion. The Assembly presents a lifestyle experience for the fun and spontaneous gentlemen who appreciate quality. Also conceptualized as a place for like-minded individuals to gather, The Assembly also has a cafe called The Assembly Ground. The identity crafted for The Assembly is one that is classic, versatile and playful. The logo mark relates to the brand name and the idea of the store and cafe as places of convocation. The Assembly's colour palette is one that is highly saturated, whilst its graphic system consists of fabric patterns.

Design agency: Bravo Creative director: Edwin Tan Designer: Jasmine Lee Client: Benjamin Barker

The Assembly 是一家多标签的男士时尚生活馆,店铺为重视品质的有趣绅士提供独特的时尚生活体验。这里也是方便人们聚会的概念化空间。店内咖啡馆名叫"The Assembly Ground",设计师为其创造了经典、多面、有趣的品牌形象。品牌标识与品牌名称、店铺理念相关联。店面采用高饱和度配色设计,平面设计方案由织物图样组成。

设计机构:Bravo 设计公司 创意总监:埃德温·谭 设计师:贾斯敏·李 委托方:本杰明·巴克

JAKARTA, INDONESIA

KOULTOURA

'We Bought a Zoo'

Koultoura 咖啡馆 / 印度尼西亚，雅加达

"我家买了动物园"

The area where the coffee shop is established is packed with students (high school and university students). This was a potential market to grab, and the designers thought creating characters would be something that will appeal to them. They wanted the brand to be more approachable in terms of how the target market can relate these characters to their own personality: the gentleman (bear), the hipster artist (rabbit), the wise guy (owl), the cool musician (fox) and the stylish dreamer (penguin). Combined with the largest potential group of market (the students), the creators were hoping that this would be a hit.

Design agency: FullFill Designer: Teddy Aang, Gerson Gilrandy, Jagat Natha Client: Koultoura Coffee

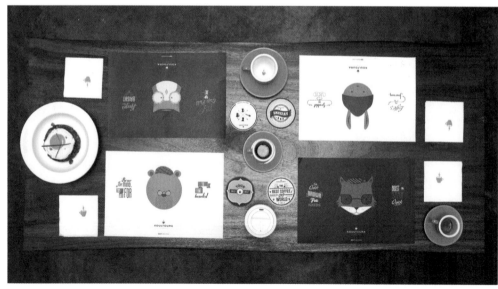

咖啡馆所在的区域有大量（高中和大学）学生出没，呈现出巨大的潜在市场，设计师认为个性化的店铺设计会对学生消费者产生较大的吸引力。品牌形象平易近人，促使目标人群能够在品牌的这些特点中找到自身共鸣：绅士（熊），时髦艺术家（兔子），聪明人（猫头鹰），酷酷的音乐家（狐狸）和时尚的梦想家（企鹅）。这样的品牌设定预计将在最大的潜在消费人群中十分流行。

设计机构：FullFill 工作室　设计师：泰迪·阿昂，格尔森·吉尔兰迪，贾加特·那斯
委托方：Koultoura 咖啡馆

JAKARTA, INDONESIA

THE PLAYROOM

Fun Illustration in a Play Venue

游戏室咖啡馆 / 印度尼西亚，雅加达

游戏空间里的趣玩插画

It's time to play! This massive 3-storey space is the perfect venue for fun and famished friends. You can play billiards, shoot some darts, watch a football match, and beat your friends at beer pong – all at The Playroom. So have fun and get fed, because you can get some grub there too!

Design agency: FullFill Designer: Teddy Aang, Astrid Prasetianti Client: Koultoura The Playroom

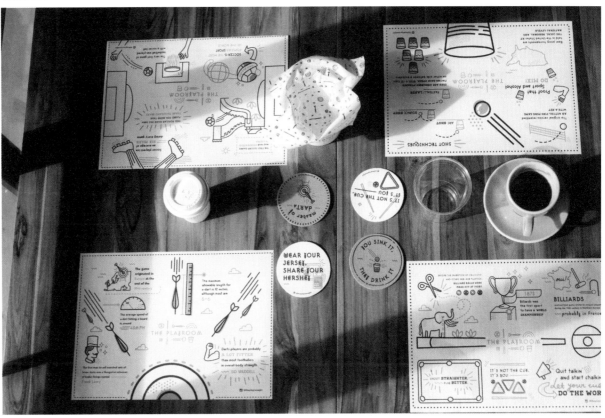

游戏时间到！这间巨大的三层建筑是朋友休闲聚会的绝佳场地。人们可以在这里玩台球、飞镖、观看足球比赛，或者玩啤酒乒乓球。快来享受一段游戏和美食时光！

设计机构：FullFill 工作室 设计师：泰迪·阿昂，阿斯特丽德·普拉瑟蒂安提 委托方：游戏室咖啡馆

MEDELLIN, COLOMBIA

9 3/4

Shh! This Is Not THE PLATFORM

9¾ 咖啡 / 哥伦比亚，麦德林市

嘘，这不是站台

9 ¾ is a bookstore cafe specialized in children, but where adults can also have some fun. PLASMA NODO believe that cities need warm and nice meeting places that welcome us and invite us to learn having fun with families and friends, sites where people not only buy but go and have a good time. The areas for children in 9 ¾ are small hiding spots or places where they can draw, rest and play while learning and enjoying a good book. For grown-ups there are private reading rooms and also tables for sharing, all surrounded by warm materials, furniture and decoration objects that speak of the joy that gives us a good story, a good book. The coffee is one of the best in town, prepared by experts and broght from the best Colombian origins. The creators believe in imagination, in magic, in dreams, in memories. The creators believe that the best ideas and conversations come easily with a good cup of coffee. They know that technology can be the magic wand to enter unimaginable worlds but it will never dethrone the King: the book.

Design agency: PLASMA NODO Designers: Daniel Mejía, Sara Ramírez, Carlos García, Felipe López, Maria F. Hormaza, Juan S. Tabares, Laura Palacio Photography: Daniel Mejía Client: 9 ¾ Bookstore + Café

9¾ 是一间主要针对儿童消费者的书店咖啡馆，成人也可以在此找到乐趣。城市需要温馨的优质聚会空间，吸引人们学会与家人、朋友共度欢乐时光。人们不仅可以在此消费，还可以玩得开心。店内供儿童玩耍的区域可以玩捉迷藏、画画、休息，或者边玩边看书。除此之外，店内专门为成人开设了私人阅读室、共享桌，均有温暖的材质、家居和装饰物包围，营造出好书给人带来愉悦的氛围。店内出售的咖啡称得上是城内品质最好的咖啡之一，从哥伦比亚购入优质原料，由专家制作完成。店主信奉想象力、魔法、梦想和回忆。也信奉优质的咖啡能够激发最好的思想和沟通。技术可以成为帮助人们进入未知世界的魔杖，但它永远无法撼动国王——书在这个世界的重要位置。

设计机构：PLASMA NODO 工作室 设计师：丹尼尔·梅希亚，莎拉·拉米雷斯，卡洛斯·加西亚，菲利普·洛佩兹，玛利亚·F·霍尔马扎，胡安·S·塔瓦雷斯，劳拉·帕拉西奥 摄影：丹尼尔·梅希亚 委托方：9¾ 书店及咖啡馆

GUATEMALA CITY, GUATEMALA

TALK' BY TALKIN' HEADS

Elegant New York Fashion

Talk'By Talkin'Heads 咖啡馆 / 危地马拉，危地马拉市

优雅的纽约时尚风

Talk is the New In-House Café by Talking Heads, the leading hair studio in Guatemala City. Talk was created to cater its VIP clients with high end gourmet drinks, entrees and main courses, while they are getting their hair, nails or makeup done. The project was commissioned by Talkin Heads to MILKnCOOKIES, the brief was to create a new brand under the Talkin Heads umbrella, that would remain linked to their TH Brand, but with an identity of its own. The Talk's Brand Personality was designed with a minimalistic approach, giving a nod to the elegance and simplicity of New York's fashion industry, yet casual enough for it's clients to relax while they are being indulged.

Design agency: MilknCookies Designer: Claudia Argueta, Gustavo Quintana Photography: Diego Castillo, Claudia Argueta Client: Talkin'Heads

本案是危地马拉城中最受欢迎的美发工作室Talkin'Heads开设的Talk店内咖啡馆。建造这间咖啡馆的目的是服务美发工作室的VIP客户,在客户享受美发、美甲和化妆服务时向他们供应高端食物与饮料。MILKnCOOKIES工作室受邀按照Talkin'Heads的要求完成这项设计,以简约的手法打造出优雅、简洁的纽约时尚风,同时兼顾客户的舒适度。

设计机构:MILKnCOOKIES工作室 设计师:克劳迪娅·阿古塔,古斯塔沃·昆塔纳 摄影:迭戈·卡斯蒂略,克劳迪娅·阿古塔 委托方:Talkin'Heads美发工作室

ABU DHABI, THE UNITED ARAB EMIRATES

BARBER & SHACK

When Barber Shop Meets Coffee Shops

Barber & Shack 咖啡馆 /
阿拉伯联合酋长国，阿布扎比

当理发店遇见咖啡馆

The concept of Barber & Shack was born due to the scarcity of quality barbers and coffee shops in Abu Dhabi. The initial plan was to start a branded barber shop which could develop into becoming a major 'line' in its industry. However, the idea later progressed into a combination of a barber shop and coffee shop in order to provide clients with considerable comfort and giving them the option of having a coffee while they shave, or having a coffee while they wait for their turn to shave.

Design agency: Cosa Nostra Creative director: Gracjan Wrzqchol Art Director: Patrycja Kordowska

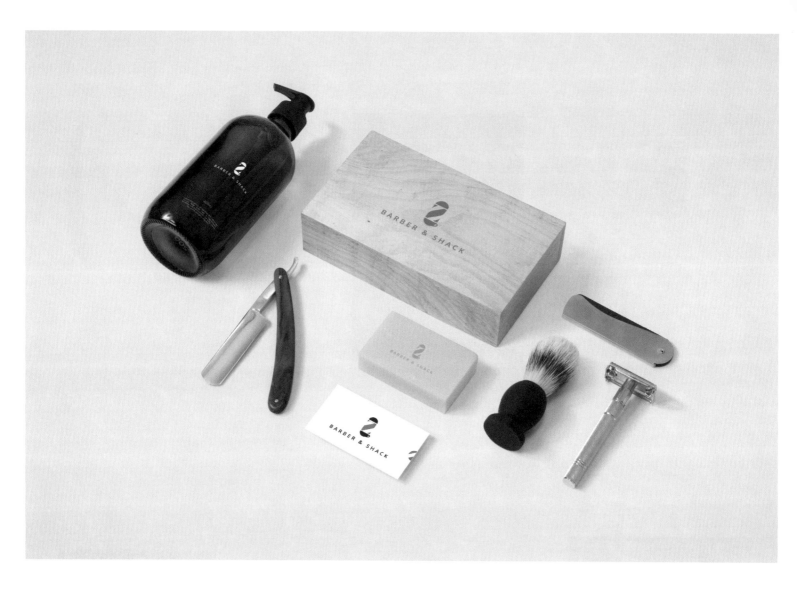
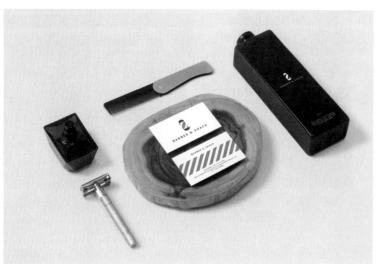

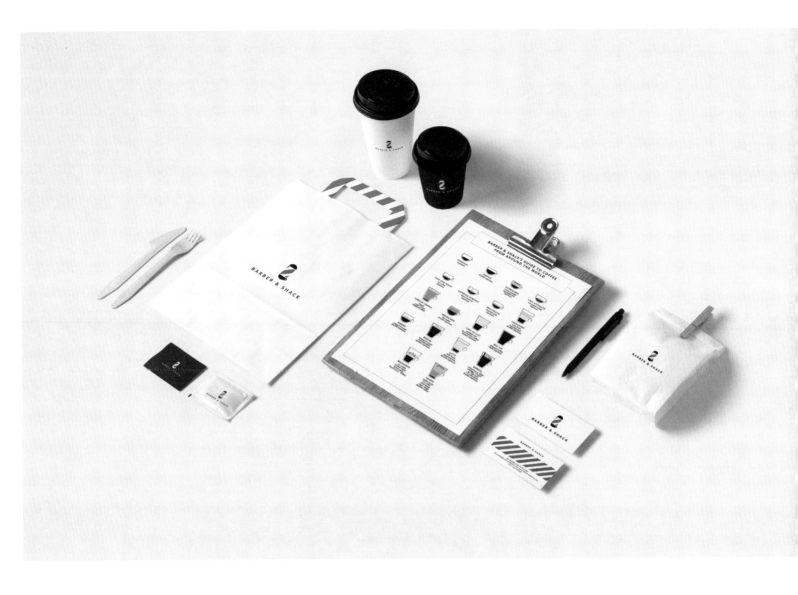

Barber & Shack 的品牌理念产生于阿布扎比对优质理发店和咖啡馆的缺乏。最初的计划是建立一个能够发展成为行业领头羊的理发品牌。然而，这一商业构想随后发展成理发店和咖啡馆的结合体，为顾客提供舒适的服务，理发的同时可以享用美味的咖啡，抑或来一杯咖啡打发等待理发的时间。

设计机构：科萨诺斯特拉工作室 创意总监：格拉经·乌拉扎霍尔
艺术指导：帕特里夏·科多斯卡

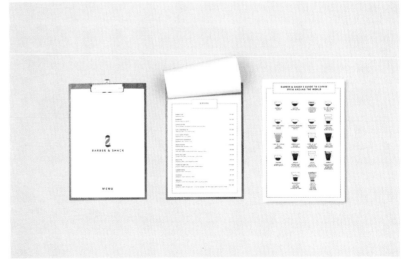

BOULDER, USA

CHE CAFE

Everyone Loves Llamas!

Che Cafe 咖啡馆 / 美国，博尔德

人人都爱羊驼！

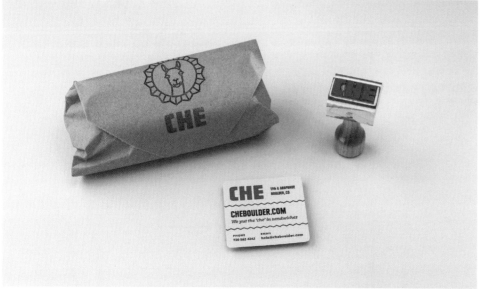

There's no doubt about it, everyone loves llamas. And when you're a restaurant with a demographic composed primarily of teenagers and young adults, a silly looking, spit hawking, long-necked mascot is definitely an asset. The word 'Che' is used as 'pal' or 'friend' in Argentina, which is where the cafe received its namesake. Creating an identity that is playful, bright, silly, and fun was a part of a conscious decision to distance the cafe from connotations with the historical figure Che Guevara. The llama, common in Argentina, is the focal point of the identity and immediately sets the tone for what visitors should expect.

Design agency: Cast Iron Design Art Director: Jonathan Black and Richard Roche Designer: Jonathan Black Illustrator: Jonathan Black Client: Che Café

毫无疑问，每个人都喜欢羊驼。在一个主要消费者是青少年和年轻人的餐馆中，长脖子、笨笨的羊驼形象无疑是一个加分项。"Che"一词在阿根廷的用法类似"pal"和"friend"，是朋友之意。设计师力求打造一个充满趣味、明快、搞笑的品牌形象，有意识地与历史人物切格瓦拉区分开来。阿根廷常见的羊驼形象构成了品牌形象的重要元素，有效地定下了整个店铺的基调。

设计机构：Cast Iron 设计公司 美术指导：乔纳森·布莱克，理查德·罗什 设计师：乔纳森·布莱克 插画师：乔纳森·布莱克 委托方：Che Café 咖啡馆

KUALA LUMPUR, MALAYSIA

THE GRUMPY CYCLIST

A Cyclist-friendly Thirdwave Café

The Grumpy Cyclist 咖啡馆 / 马来西亚，吉隆坡

骑行者的异度空间

Branding for The Grumpy Cyclist, a cyclist-friendly thirdwave coffee joint in Kuala Lumpur, Malaysia. With the logotype as its main axis, different emblems are to be introduced at different events and workshops. The interior work is by Ard Krate.

Designer: Koyoox Interior design: Abdul Rahman Aljunid, Ard Krate Photography: Koyoox Client: The Grumpy Cyclist

本案是为马来西亚吉隆坡的 The Grumpy Cyclist 进行的品牌形象设计。The Grumpy Cyclist 是一家针对自行车骑行者的咖啡连锁店。从品牌商标延伸出的不同设计应用于各种场合和工坊活动中。室内设计由阿尔德·柯莱特完成。

设计师：Koyoox 室内设计师：阿卜杜勒·拉赫曼·阿尔朱尼，阿尔德·柯莱特 摄影：Koyoox 委托方：The Grumpy Cyclist 咖啡馆

CALIFORNIA, USA

LE PASTEL

Pet Dog Inspiration

宠物乐园咖啡馆 / 美国，加利福尼亚

狗狗带来的灵感

The title of this project, 'PET GROUND', is the combination of two words PET and PLAYGROUND. Inspired by the affection towards dogs, Jeong designed a concept of a virtual pet cafe in which owners enjoy their food and drinks while their pets enjoy playtime with their friends. Hence, this place is great for pets to socialize.

Designer: Jeong Min Kim Client: PETGROUND Cafe

本案中的项目名"petground"是宠物和操场两个词合成而来的。设计师在自己对狗的喜爱中找到设计灵感，创造出宠物咖啡馆的概念，宠物主人可以将宠物带到咖啡馆与其他小动物玩耍互动，自己则享用美味的食物和饮品。因此这里是宠物获得社交的理想去处。

设计师：金正民 委托方：宠物乐园咖啡店

SYDNEY, AUSTRALIA
THE LOCAL MBASSY

A Reflection of the Australian 1920s

THE LOCAL MBASSY 咖啡馆 / 澳大利亚，悉尼

二十世纪二十年代澳式风情

Creative director and designer Korolos Ibrahim was deeply enthused by the Australian prohibition-era. 'We're taking off our hats to the local hooligans, the revolutionaries and to those who made a difference in moulding up today's Australian art, fashion and coffee culture', says Korolos. Upon receiving the brief for the undeveloped start-up business, it was fundamental that a full creative direction, business development, concept, brand and design were developed. The boiler room inspired interior comes with rigueur exposed beam and bulbs by Antique collector and designer David Haines; raw concrete finishes, one off refurbished furniture and a larger-than-life feature mural painted by local Australian street artist, Sid Tapia. Korolos hopes to provide coffee-swilling jazz aficionados and foodie enthusiasts a destination café that combines good food, positive vibes and a whole lot of good art.

Design agency: Korolos Design Designer: Korolos Ibrahim Artist/muralist: Sidney Tapia
Photography: Shayben Moussa Fit-out pipe Installation: David Haines Client: Marcus Gorge

身兼创意总监和设计师的克罗洛斯·易卜拉欣对澳大利亚的禁酒令时期十分着迷。"我们希望向当地革命者以及那些塑造了现代澳式艺术、时尚和咖啡文化的人致敬。"设计师这样说道。当他接到这份待开发的商业企划时，设计师需要创造出一套创意导向、业务发展、整体理念、品牌和设计。以锅炉房为灵感来源的室内设计方案将房梁暴露在外，使用古董收藏家、设计师大卫·海恩斯的灯泡产品；裸露混凝土饰面，翻新家具，澳大利亚本土艺术家西德尼·塔皮亚创作的超大壁画。设计师希望为喜爱咖啡的爵士爱好者和美食爱好者提供一个能够同时享受到美味食物、优美音乐和优质艺术的理想去处。

设计机构：克罗洛斯设计工作室 设计师：克罗洛斯·易卜拉欣 美术/壁画：西德尼·塔皮亚 摄影：沙依本·穆萨 Fit-out pipe 装置：大卫·海恩斯 委托方：马库斯·戈尔热

LONDON, UK

DISHOOM

Inspiration from Beautiful Old Irani Cafés

DISHOOM 咖啡馆 / 英国，伦敦

灵感源于美丽的老式伊朗咖啡馆

Dishoom take inspiration from Bombay's beautiful old Irani cafés. & SMITH has been working with them over the last two years to evolve their visual identity and help with anything print based.

Design agency: & SMITH Designer: Sam Kang Photography: & SMITH

Dishoom 咖啡馆的设计灵感源自孟买美丽的老式伊朗咖啡馆。在过去的两年中，& SMITH 设计公司一直致力于这个项目，力求创造出更好的视觉形象，同时协助完善平面设计。

设计机构：& SMITH 设计公司 设计师：山姆·康 摄影：& SMITH 设计公司

YEREVAN, ARMENIA

GAWATT COFFEE SHOP

Graffiti in Steampunk Style

GAWATT 咖啡馆 / 亚美尼亚，埃里温

蒸汽朋克风格的涂鸦

Gawatt, which in Armenian means 'cup', not just an ordinary cup. It is take-out coffee shop in steampunk style. Without first two letters it is a name of the universal steam engine inventor. Backbone designed different sizes of cups depending on the power charge of their contained-watt, kilowatt, megawatt, terawatt.

Design agency: Backbone Branding Studio Art director: Stepan Azaryan Designer: Karen Gevorgyan Photography: Backbone Branding Studio Client: Artur Danielian

"GAWATT"在亚美尼亚语中是杯子的意思,但不是普通的杯子,它指的是蒸汽朋克风格的外带咖啡杯。去掉前两个字母就是大名鼎鼎的蒸汽机发明者的名字。设计师从电力单位汲取灵感设计了不同尺寸的杯子。

设计机构:Backbone 品牌设计公司 艺术总监:斯特潘·阿扎严 设计师:凯伦·盖沃谙 摄影:Backbone 品牌设计公司 委托方:阿图尔·丹尼利恩

LONDON, UK

VILLANDRY

Inspirations from Traditional Brasserie Menu Frames

维朗德里 / 英国，伦敦

传统啤酒铺菜单给予的灵感

Villandry is an all-day French restaurant, cafe, foodstore, bakery and bar located in London and Bicester Village. Mind Design developed the identity, packaging, signage system and all printed materials. The identity is based on a set of five shapes which refer to traditional brasserie menu frames. The shapes are either used individually or combined to create a pattern. A bright colour scheme was chosen in contrast to the classic shapes.

Design agency: Mind Design Client: Villandry

Villandry 是坐落在伦敦和比斯特村的全天营业的法式餐厅、咖啡馆、烘焙坊和酒吧。Mind Design 为其进行了品牌形象、包装、标识系统和所有印刷材料的设计工作。品牌形象以一组 5 个图形为基础，灵感来自传统啤酒铺菜单设计。这些图形可以单独使用或组合搭配。设计中使用了明快的配色，与经典的图样形成反差。

设计机构：Mind 设计公司 委托方：维朗德里餐厅

Create an identity for the interior of Vintage and Coffee for Music that represents the characteristics of the brand through different communication tools. A corporate image applied to the corporate space. An identity that represents the key values of the brand is created and particularly reflected through the naming and slogan. Fresh colours exchanged according to their communicative use present a world full of new sensations and tastes.

Design agency: +Quespacio Designer: Ana Milena Hernández Palacios
Photography: David Rodríguez Pastor

通过不同的沟通工具展现品牌特征。在整个店面空间使用统一的企业形象，反映品牌的核心价值，特别体现在命名和标语上。不同的沟通功能中使用不同的鲜明配色，展现一个充满新感官和品味的世界。

设计机构：+Quespacio 工作室 设计师：安娜·米莱娜·埃尔南德斯·帕拉西奥斯 摄影：大卫·罗德里格斯·帕斯托

BARCELONA, SPAIN

VINTAGE AND COFEE FOR MUSIC

A World full of New Sensations and Tastes in Fresh Colours

音乐复古咖啡馆 / 西班牙，巴塞罗那

鲜明配色下的新感官

Cafe Renzo is attached to the Astrup Fearnley Museum of Modern Art at Tjuvholmen in Oslo. The building was designed by Renzo Piano, who also designed the café's inventory. Alongside with the café's interior & surroundings, the objective was to honor Renzo Piano's architecture, the employees' work for Cafe Renzo and Astrup Fearnley's Modern Art.

Design agency: Pocket Designer: Nicklas Haslestad Photography: Nicklas Haslestad, Stian Broch Client: Fursetgruppen AS

伦佐咖啡馆位于奥斯陆 Tjuvholmen 的阿斯楚普费恩利现代艺术博物馆。咖啡馆所在的建筑和室内设计均由伦佐·皮亚诺完成。除了咖啡馆的室内及整体空间设计，项目还着重纪念伦佐·皮亚诺的建筑作品、员工们为伦佐咖啡馆付出的辛勤劳动以及阿斯楚普费恩利现代艺术博物馆。

设计机构：Pocket 工作室 设计师：尼克拉斯·海思乐 摄影：尼克拉斯·海思乐，什蒂安·布罗赫 委托方：Fursetgruppen AS 公司

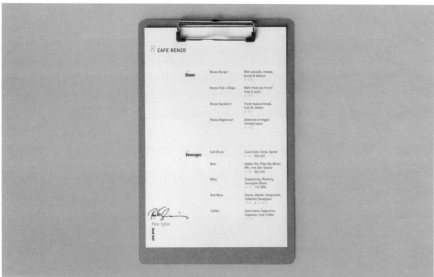
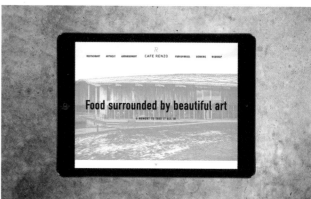

OSLO, NORWAY

CAFE RENZO BY RENZO PIANO

Visual Identity of Architectural Thinking

伦佐·皮亚诺的伦佐咖啡馆 / 挪威，奥斯陆

富有建筑设计感的视觉形象

JAZZ HOUSE CAFE

BAKU, AZERBAIJAN

Jazz in Sketches

Jazz House 咖啡馆 / 阿塞拜疆，巴库

速写中的慵懒

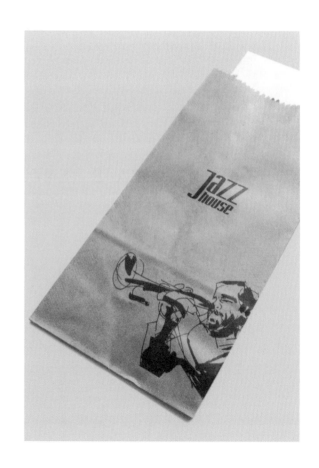

A small café has been opened in the center of the city. Its main function was to join all jazz-lovers. The branding includes: logo, stationery (paper bag, blanks, cups, mugs, visit cards, under-plate papers, envelope), posters, and sign board. Usually in all jazz-concerned logotypes musical instruments were used. The designer decided to use typography to make the design plain and easy to percept.

Designer: Aysel Sadigova　Client: Jazz House Cafe

本案的主体是位于市中心的一家小咖啡馆。其主要功能是汇集所有的爵士乐爱好者。品牌形象设计内容包括：品牌标识、文具（纸袋、空白表格、茶杯、马克杯、卡片、餐垫纸、信封）、海报和标志板。所有与爵士乐有关的标志设计中通常都会用到乐器的图样。设计师选择反其道而行之，打造简单、易辨识的设计。

设计师：艾塞尔·萨迪戈娃　委托方：Jazz House 咖啡馆

NIZHNY NOVGOROD, RUSSIA
MUSICAL CAFE 'ROYAL'

The Colours of the Royal: Black & Gold

"皇家"音乐咖啡馆 / 俄罗斯，下诺夫哥罗德

最华丽的色彩：黑与金

It all started with the development of the sign for a music cafe 'Royal'. Following the familiar icons, a full corporate identity was further developed. The concept is: the Russian word meaning a piano keyboard musical instrument stands, at the time, as the English meaning of the word Royal - meaning kingdom, gold and pretentiousness. It is on this game transfers and built visual concept logo. Cafe 'Royal' is a good place to listen to live music, eat and drink.

Design agency: Chrome studio Designer: Grigory Khromov
Client: Individual entrepreneur Photography: Grigory Khromov

本案是为名为"皇家"的音乐咖啡馆进行的标识设计。图标设计完成后,进一步开发了全套的企业形象。设计概念如下:正如英语词汇"Royal"有王国的含义,俄语中意为钢琴键盘的词汇也代表金色和气派。在此基础上,设计师创造了品牌标识的视觉概念。"皇家"咖啡馆无疑是一个欣赏现场音乐、美食和美酒的好去处。

设计机构:Chrome 工作室 设计师:格里戈里·赫罗莫夫 摄影:格里戈里·赫罗莫夫 委托方:独立企业

BUDAPEST, HUNGARY

KINO CINEMA & ART CAFE

Design with Images

吉纳影院和艺术咖啡馆 / 匈牙利，布达佩斯

设计的影像

Identity and package design for KINO, a really friendly café place with a cinema in the heart of Budapest. Graduation project at the Hungarian University of Fine Arts.

Designer: Szani Mészáros Client: KINO Photography: Judit Kozma, Szani Mészáros

本案是为布达佩斯市中心一家温馨的影院兼咖啡馆进行的品牌形象设计。是匈牙利美术大学一位学生的毕业设计项目。

设计师：赞妮·美莎露丝 摄影：尤迪特·科兹马，赞妮·美莎露丝 委托方：吉纳影院和艺术咖啡馆

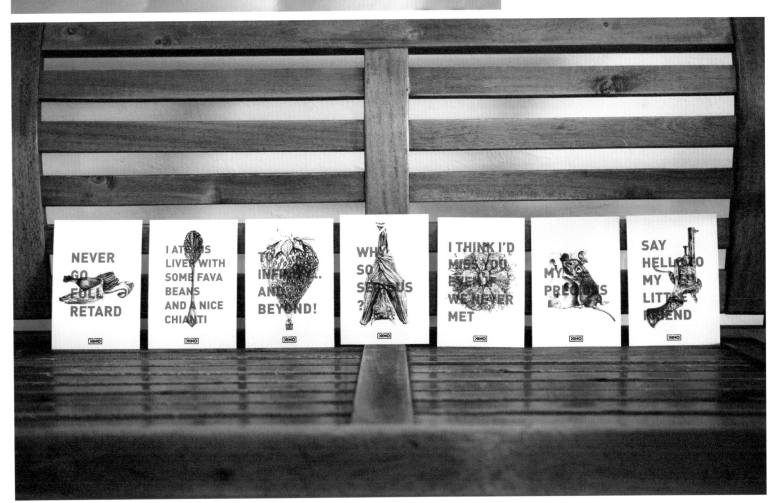

INDEPENDENT COFFEE SHOP

独立咖啡馆

GUADALAJARA, MEXICO

PERSILLADE-EAST MELBOURNE

Multi-functions of a Café

东墨尔本——Persillade 咖啡馆 / 墨西哥，瓜达拉哈拉

咖啡馆的多功能性

Persillade — aptly named after the zesty French seasoning — is a welcome addition to the exclusive neighbourhood of East Melbourne. Owned by Aidan and Tanya Raftery, Persillade is a bustling cafe by day and elegant French bistro by night. Working collaboratively with Aidan and Tanya, Clear designed the identity, signage and menus as well as the interior materials and styling. Textural, layered, personable and refined contemporary dining experience.

Design agency: Clear Designer: Matthew McCarthy, Zoran Konjarski Photography: Scottie Cameron Client: Persillade

Persillade 以法国香料命名，是东墨尔本很受欢迎的一家店铺。Persillade 归艾登·拉夫特里和塔尼娅·拉夫特里所有，白天是熙熙攘攘的咖啡馆，夜晚化身优雅的法式小酒吧。Clear 工作室与艾登·拉夫特里和塔尼娅·拉夫特里一同完成了品牌形象、标识、菜单以及室内和整体风格的设计工作。店内提供有质感有层次的个人化设计下的精致的现代就餐体验。

设计机构：Clear 工作室 计师：马修·麦卡锡，佐兰·康亚斯基 摄影：斯科蒂·卡梅伦 委托方：Persillade 咖啡馆

NEW YORK CITY, USA

MAMAN NYC

'I was inspired by the waves of the sea'

马曼咖啡馆 / 美国，纽约

"我在海浪中找到了灵感"

Maman's curious, minimalist logo heralds this modern sensibility. Responding to Maman's brief asking for handwritten typography, Roux included a design departure as a third study. Roux visited New York for the first time just before the bakery opened and was pleased to see her logo deployed throughout the space. 'In the way they did their cards, tags, their signage on the wood—all those great details—Elisa and Benjamin took my sketch and made it special, something precious that you want to hold on to.'

Designer: Alexia ROUX Client: Maman

好奇、极简的品牌标识透露着马曼咖啡馆现代而感性的一面。应品牌要求,设计师专门设计了手写字体和标识。此外,设计师还在店铺开张前特意赶赴纽约,实地检验设计成果。"我的设计在店铺的卡片、标签、木质招牌等细节上得到应用,似乎变得更为特别,成为了具有真贵价值的东西。"

设计师:亚莉克希亚·鲁 委托方:马曼咖啡馆

MELBOURNE, AUSTRALIA
GRASSHOPPER

A Café with Nature in It

蚱蜢咖啡 / 澳大利亚，墨尔本

咖啡馆里的自然世界

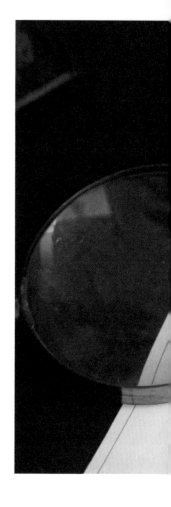

Grasshopper Cafe and Bar is a self-authored young, hip cafe in an urban setting. Inspired by the Michael Pollan quote, 'Eat food. Not too much. Mostly plants', the restaurant offers food that makes you feel good. Its aesthetic is rooted in vintage field guides and entomology. The restaurant has a direct connection to local farmers, bakeries, and butchers to make sure all its ingredients are as fresh and direct as possible. Its menu is seasonally updated, including an ingredient index so the restaurant develops a completely transparent, honest, and trusted relationship with its customers and community. Grasshopper is a place to celebrate healthy living and enjoying life amongst friends and family.

Designer: Lacy Kuhn Photography: Lacy Kuhn

蚱蜢咖啡是墨尔本市内一间年轻、时髦的自营咖啡馆。品牌灵感源自迈克尔·波伦的名言"吃食物。植物为主。别太多。"店内供应让人感觉舒服的食品。设计以复古的野外指南和昆虫学。咖啡馆与当地农户、烘焙坊和肉铺建立直接联系，确保所有原料都尽可能新鲜。店内供应菜品随季节调整，内容包含原料表，使咖啡馆与顾客以及社区之间建立完全透明、诚实、可信的关系。蚱蜢咖啡乐享健康生活，是与亲朋好友欢度美好时光的理想去处。

设计机构：莱西·库恩工作室 设计师：莱西·库恩 摄影：莱西·库恩

MOSCOW, RUSSIA

DRUZHBA CAFE

The Image of the Bicycle 'Tandem'

德鲁日巴咖啡 / 俄罗斯，莫斯科

"tandem" 的自行车形象

Cafe 'Druzhba' was conceived as a meeting place for friends and acquaintances, democratic cozy place to spend time together. The image of the bicycle 'tandem' came while working on the interior, where the designers have added it in the space of a cafe. Here it is! The image of friendship, the image of unity and movement, the image of the modern lifestyle - healthy and positive.

Design agency: Panfilov&Yushko Creative Group Designer: Dmitry Panfilov Photography: Daria Doroshenko Client: Druzhba Cafe

德鲁日巴咖啡是一个与友人聚会的理想场所,一同度过惬意的休闲时光。名为"tandem"的自行车形象是室内设计过程中创造出来的,随后设计师将它用在咖啡馆的整体设计中。它代表了友好、统一和动感,以及现代生活方式——充满健康和积极的力量。

设计机构:潘菲洛夫与尤什科创意工作室 设计师:德米特里·潘菲洛夫 摄影:达莉亚·多罗申科娃 委托方:德鲁日巴咖啡

CHANTHABURI, THAILAND

C.A.P.

The Renovation of a Historic Café

C.A.P. 咖啡馆 / 泰国，尖竹汶

老咖啡馆的新生

Café and People (C.A.P.) is a coffee shop, which its long-story was originated in the historic community of Chanthaburi province, Thailand. Its logo was designed based on the various popular and distinguished characteristics and products of Chanthaburi such as rabbit, waterfall, sapphire, mat etc. These key elements are represented into one Chanthaburi's symbol and as portrayal of the community. This project was initiated under the intention to modernize the logo, and establish new branding image as an inspiration to ultimately transform community's perception of the existing logo.

Design agency: Wide And Narrow Co., Ltd. Designer: Wide And Narrow Co., Ltd. Client: C.A.P.

咖啡与人（C.A.P.）是一间历史悠久的咖啡馆，坐落在泰国尖竹汶府的传统社区中。店铺的品牌标识设计灵感源自兔子、瀑布、蓝宝石等多种在当地流行和有代表性的特色与产品。设计师将这些主要元素整合成尖竹汶以及当地社区的象征。发起本项目的意图是将标识现代化升级，建立新的品牌形象，进而激发人们做出改变，最终改变社区对原有标识的认识。

设计机构：Wide And Narrow 有限公司　设计师：Wide And Narrow 有限公司　委托方：C.A.P. 咖啡馆

SFAX, TUNISIA

THE CUP

Modern Vintage Coffee Style from Tunisia

杯子咖啡馆 / 突尼斯，斯法克斯市

来自突尼斯的复古咖啡风格

Branding and Pattern design for a coffee shop. The use of different textures and papers in this visual identity were adopted to strengthen the concept via modern vintage coffee style.

Design agency: agence 360 degrees Client: The cup

本案是为一家咖啡馆进行的品牌形象与平面设计。视觉形象设计中使用了不同的材质与纸品，通过现代的复古咖啡风格强化了店铺理念。

设计机构：360度工作室 委托方：杯子咖啡馆

SFAX, TUNISIA

LA MASCOTTE

Abstract, Minimal Perspective to a Modern Experience

马斯科特咖啡馆 / 突尼斯，斯法克斯市

模糊化的抽象设计理念

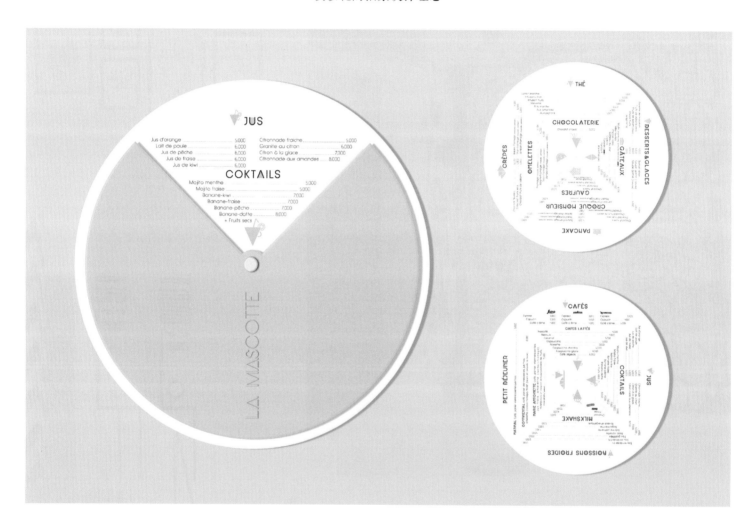

Visual identity for a modern coffee shop. The concept is based on an abstract, minimal perspective to link with the modern experience that 'La Mascotte' offers to their client.

Design agency: agence 360 degrees   Client: La Mascotte

本案是为一家咖啡馆进行的视觉形象设计。设计师采用了抽象的设计理念，将马斯科特咖啡馆提供给顾客的现代式休闲就餐体验模糊化。

设计机构：360度工作室 委托方：马斯科特咖啡馆

TAIWAN, CHINA

MICIO CAFFE

The Geometric Patterns for a Cat

米球咖啡 / 中国，台湾

猫的几何图形

Micio means kitten in Italian while the name and persistence of the brand come from a kitten with tenacious vitality. The founder of Micio Caffe wishes to condense this strength into the brand philosophy and the perseverance of food quality. Transform Design therefore create a space of lovely cat images and lots of carefree playing via straightforward logo design. Simple geometric patterns are employed in the design to express a carefree experience and resilient attitude.

Design agency: Transform Design Creative director: Leo Huang Designer: Celine Shen Client: MICIO CAFFE

米球 MICIO 是意大利语小猫的意思，品牌的名称与坚持都来自于一只拥有坚韧生命力的小猫。MICIO CAFFE 的创办人希望能将这份力量转化至品牌的精神，以及对食物品质的坚持。因此我们希望能够将猫的形象与让猫轻松玩耍的空间，透过标志简单而直接的带入品牌之中。造型采用最基本的几何方式，试着用最纯粹的方法，将柔软、无压力的感受转化为坚持而有力的态度。

设计机构：瑜悦设计 创意总监：黄国瑜 设计师：沈思伶 委托方：米球咖啡

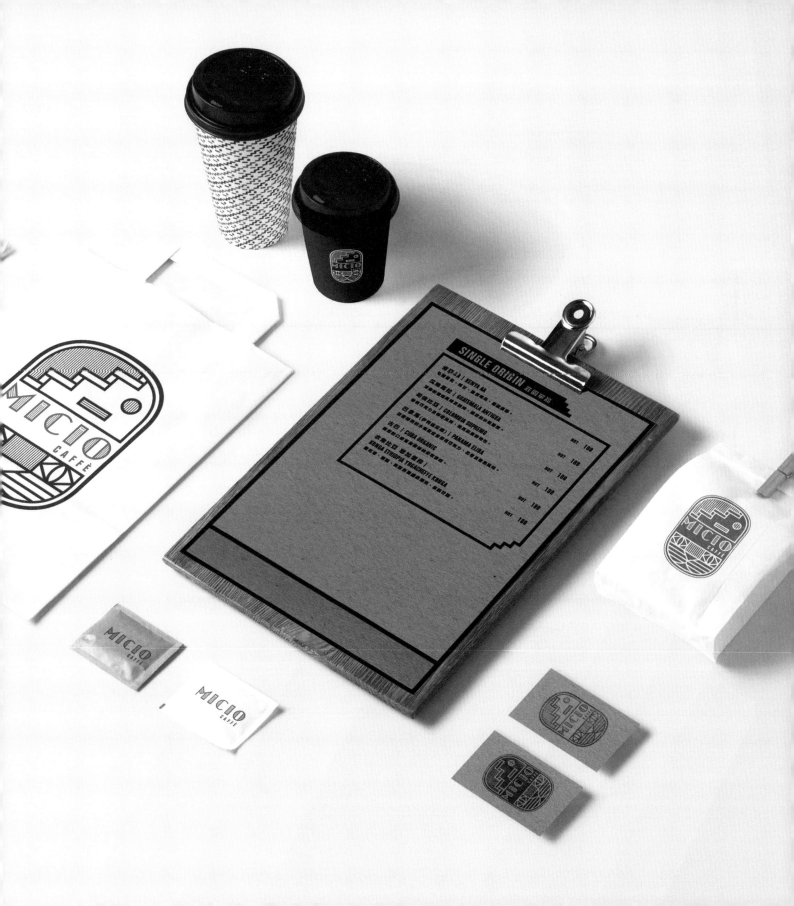

SALZBURG, AUSTRIA

WOHNZIMMA

Vintage Sealed in Rubber Stamps

WohnZimma 咖啡馆 / 奥地利，萨尔斯堡

橡皮图章象征的复古氛围

WohnZimma is a recently opened coffee shop located in a small village in Salzburg, Austria. The basic idea of the café is to serve homemade cakes and selected regional products. To highlight the handmade character, a rubber stamp is used on all corporate documents, which also reflects the interior and gives a special vintage touch to the whole brand appearance.

Design agency: Miriam Weiss Photography: Miriam Weiss
Client: Andrea Minichberger | owner of the coffee shop

WohnZimma 是奥地利萨尔斯堡一个小村庄上新开业的咖啡馆。店铺的基本理念是供应自制糕点和精选的当地产品。为了突出手工自制的特点，所有的公司文件上都使用了橡皮图章形象。室内设计中也采用了这一元素，为整个品牌形象营造出独特的复古氛围。

设计机构：米利安·韦斯 摄影：米利安·韦斯 委托方：安德里亚·米尼希伯格（咖啡馆店主）

HERTFORD, UK

FOUNDATION COFFEE HERTFORD

Kraft Paper and Brand Minimalism

赫特福德基金咖啡 / 英国，赫特福德郡

牛皮纸色彩下的简约风格

Foundation Coffee was a new independent coffee shop in Hertford. The clients for this wanted Ryan to create the initial branding; this included the brand identity and various other products. The client wanted to keep the brand minimalist.

Design agency: Ryan Gaston Designs Ltd. Designs Designer: Ryan Gaston
Photography: Ryan Gaston

基金咖啡是赫特福德一家新近开张的独立咖啡馆。委托方希望设计师创造一个初始品牌形象，包括品牌整体形象和多种产品形象。委托人还希望设计方案体现简约的风格。

设计机构：Ryan gaston 设计公司　设计师：瑞安·加斯顿　摄影：瑞安·加斯顿

BRINDISI, ITALY

UMAMI ZEN CAFE

Visual Clash of Eastern and Western Cultures

屋纳米禅意咖啡 / 意大利，布林迪西

东西文化的视觉碰撞

Umami Zen Cafè's philosophy is to mix the Italian tradition of coffee to the typical Japanese atmosphere, representing a bridge between eastern and western culture. The originality of the idea consists in contrasts and unusual approaches, keeping however a clean and minimal image.

Designer: Daniele 'Donnie' D'Addario   Client: Umami Zen Cafè

屋纳米禅意咖啡秉承意大利传统咖啡与经典日式装潢有机融合的经营理念，代表连接东西文化的桥梁。这一构想的创意采用反差和不同寻常的方法，呈现出简洁的品牌形象。

设计师：达妮埃莱·唐尼·达达里奥   委托方：屋纳米禅意咖啡

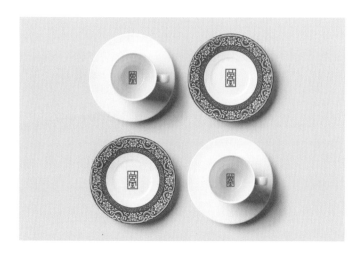

BRANDED COFFEE SHOP

品牌咖啡馆

MOSCOW, RUSSIA

BABETTA CAFE

Vintage with Memorable Details

芭贝特咖啡馆 / 俄罗斯，莫斯科

乱入的复古时尚感

Precise design filled with memorable details make it a favourite spot around Moscow. The design makes customers feel like Babetta Café has been around and popular forever: vintage 60s chairs, restored fixtures, lamps from an abandoned factory in LA, and bar constructed from welded metal pipes combine in contrast to create a friendly and cozy atmosphere equally suited for a big group of friends enjoying a fun time together or for a relaxed meal with beloved parents. A huge wood-burning stove is the heart of the restaurant and always lit. Visible from the street it signals to everyone passing by where to find the most delicious pizza in town!

Design agency: Bureau Bumblebee Creative director: Alina Pimkina Production Director: Olesya Shebetova Senior Architect: ilya Mozgunov Designers: Andrej Barmalej, Mikhail Saprykin

项目的精确设计和细节处理使其成为莫斯科一处最受喜爱的地方。设计方案让顾客觉得芭贝特咖啡馆似乎已经存在了很久：复古的20世纪60年代风格沙发，修复的旧式家具，从洛杉矶一个废旧工厂回收的灯具，以及金属管焊接而成的吧台，一同营造出友好、舒适的氛围，既适合朋友聚会也适应与家人共进美食。巨大的炉子位于餐厅中央，总是燃烧着。店铺的标识极易辨识，招徕着来往的行人进店品尝莫斯科最美味的比萨。

设计机构：大黄蜂设计工作室 创意总监：艾琳娜·品奇娜 产品总监：奥列西亚·社贝托娃 高级建筑师：伊利娅·莫兹古诺夫 设计师：安德烈·巴尔马列，米哈伊尔·萨普雷金

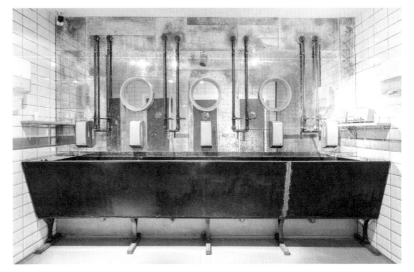

LONDON, UK

COFFEE HOUSE LONDON

Golden Embossing for London Culture

伦敦咖啡馆 / 英国，伦敦

金色雕花打造伦敦品质

Launching a new coffee brand in today's very competitive market is hugely challenging. You have to offer something truly unique, of the highest quality, along with great atmosphere. You really have to stand out in a crowd. London is a city deeply rooted in its traditions, history and architecture. Loyalties are formed in childhood and honoured for a lifetime. So the task is not just to show the outstanding benefits of the product but to weave these assets into the larger culture and themes of London culture, combining the heritage of coffee drinks with the distinctive, one-of-a-kind pleasures of London House coffees. Reynolds and Reyner tried to use mostly natural materials, golden embossing and paper with different textures.

Design agency: Reynolds and Reyner Designer: Alexander Andreyev, Artyom Kulik Client: Coffee House London

在眼下竞争异常激烈的市场中建立全新的咖啡品牌无疑是一个巨大的挑战。店家需要为顾客提供真正独特的体验：最优质的产品配合怡人的消费环境。经营这样的品牌需要脱颖而出，才能生存下来。伦敦是一个深深扎根于传统、历史和建筑的城市，人们在几时树立忠诚的概念，并用一生去实践。我们的任务不只是展现自家产品的优秀品质，也要将这些品质融入更大的伦敦文化主题之中，将咖啡的文化精髓和伦敦咖啡馆独特的品牌理念结合在一起。我们尝试使用了最天然的材料、金色雕花和不同质感的纸品。

设计机构：Reynolds and Reyner 工作室　设计师：亚历山大·安德烈耶夫，阿尔乔姆·库利克　委托方：伦敦咖啡馆

CHIHUAHUA, MEXICO

HECHO CON AMOR

Nostalgia from Grandmother's Kitchens

Hecho Con Amor 咖啡馆 / 墨西哥，奇瓦瓦

如祖母厨房般的亲切怀旧风潮

This café is conceived to be the place in Chihuahua city that brings back the nostalgia from grandmother's kitchen. Folklore developed a minimalistic concept that demonstrates the idea of the place in a very simple way. Having both concepts in mind, the artisanal and the modern, they elaborated the branding in a very traditional way, to get the perfect look and feel for the place. Folklore selected contemporary furniture to match the rest of the decoration, in which a vertical garden was included.

Design agency: Folklore Designers: Germán Aguirre and Adriana De La Torre Photography: Raúl Ramirez 'Kigra' Client: Hecho con Amor Local Coffee shop.

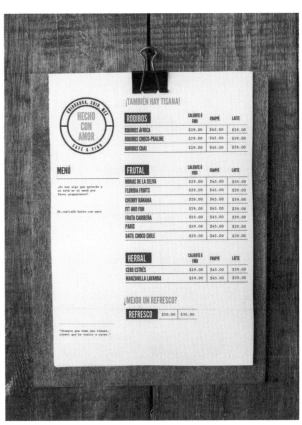

本案中的咖啡馆为奇瓦瓦市带来如祖母的厨房般亲切的怀旧风潮。设计师开发了简约风格的设计理念，以极其简单的方式展示这一空间概念。设计师以非常传统的方式将手工和现代两个概念融合在一起，创造出外观完美的品牌形象和氛围一流的店面。设计师选用了现代家具搭配店内的垂直花园等其他装饰元素。

设计机构：Folklore 工作室　设计师：热尔曼·阿吉雷，阿德里亚娜·戴拉托瑞　摄影：劳尔·拉米雷斯"Kigra"　委托方：Hecho con Amor 咖啡馆

NEW YORK, USA

LAUGHING MAN COFFEE

"All Be Happy"

LAUGHING MAN 咖啡 / 美国，纽约

"一切皆快乐"

Laughing man is Hugh Jackman's charitable organization. The brief was to create a brand that broke the mold of ethically responsible products being packaged in a rustic, hessian and often ugly way so that the products should look at home on the shelves of a luxury food hall. A silver and grey colour palette and clean typography were used, while the custom chalk type graphics used across the brand connote a hand-finished barista quality. The unique laughing face stickers on each product were designed to reflect the universal nature of laughter, goodness and ethicalmindedness, fundamental to the laughing man brand. The concept of constantly changing stickers also invited the consumer to interact with the brand directly by uploading their own laughing face on the website, for a chance to be used on future packaging.

Designer: Establishd Client: Laughing Man Coffee

LAUGHING MAN 咖啡是影星休·杰克曼创办的慈善组织。项目要求设计师打破传统的产品包装模式，塑造符合豪华食品展示空间的全新产品形象。设计师选用银色与灰色搭配简洁的排版，特制的粉笔字体图案应用在整个品牌形象中，呼应店内咖啡手工制作的品质与特点。每个产品上独特的笑脸粘贴设计传递品牌的核心理念：笑容、善良与道德。不断更换粘贴的想法鼓励顾客通过互联网上传自己的笑脸，参与未来的包装制作，与品牌直接互动。

设计机构：Establishd 工作室 委托方：Laughing Man 咖啡

MELBOURNE, AUSTRALIA

THE CORNER

A Project of Twenty-Year-Plus Life Expectancy

街角咖啡馆 / 澳大利亚，墨尔本

设计方案的保质期：20 年

The brief was to be part of a program to challenge McDonalds. This is a Future Forward project whose purpose is to stretch the brand beyond its comfort zone, to reach new customers and position the brand in the 'fresh & fast food market'. Potentially, some of the more successful elements may be integrated into other McCafé stores, enriching and up-scaling the existing food and beverage offer. Landini's approach was to design a great local café, not an evolution of McCafe. Working with such a global brand thus required a determinedly blinkered approach, one that looked to the local neighbourhood and not the past. Landini seek the 'classic' over the fashionable, the quiet over the loud. This is their approach to sustainability and many of their project's life expectancy is twenty years plus, exceeding the norm by a multiple of three. Their job was to support this with an environment that made the food the hero and to help challenge the publics and staff's perception of this brand by creating a local and unexpected solution. The brief extended to the naming, graphics, packaging, uniforms and interiors. This innovation was an immediate success, loved equally by the staff and customers.

Design agency: Landini Associates Photography: Sharrin Rees Client: McCafe

本案所属项目志在对麦当劳发出挑战。这是一项名为"未来向前冲"的工程，目的是挑战品牌跳出安全区域，尝试吸引新的顾客群，在"新鲜食品与快餐市场"中立足。一些较为成功的元素可以整合到麦咖啡馆中，丰富并发展现有的食品和饮品种类。设计师希望打造一个优秀的当地咖啡品牌，而不只是麦咖啡的升级版。与如此出色的国际品牌合作需要设计师明确工作方法，关注当地现况而不是过去的情况。在设计师的目标中，经典高于流行，静谧优于喧闹，由此实现可持续的设计。项目的预期使用期限是20多年，远远超过了普通项目的使用寿命。设计师的任务是提供环境支持，使食物成为焦点，用出人意料的设计挑战大众和员工对品牌的认识。项目要求包含了命名、图形、包装、制服和室内设计。创新之处立即获得了成功，受到员工和顾客的共同喜爱。

设计机构：兰迪尼设计公司 摄影：沙林·里斯 委托方：麦咖啡

AUCKLAND, NEW ZEALAND

MR TOMS

Simple Urban Lifestyle

Mr Toms 咖啡 / 新西兰，奥克兰

大隐于市的独居格调

Classic and engaging, the Mr Toms brand had to personify quality and old fashioned hospitality- the core values of this new venture. Fuman created a slick word mark to reflect the brands stylish and trustworthy nature with complimenting illustrations that convey a sense of the vibrancy and fun that was also paramount to the brand's identity.

Design agency: **Fuman Design Studio** Designer: **Jon Chapman-Smith** Photography: **Jon Chapman-Smith** Client: **Mr Toms**

Mr Toms 咖啡馆是一间经典而迷人的咖啡馆，优良的品质和传统的待客方式是店铺秉承的核心价值。孚曼设计工作室为品牌创造了巧妙的标语，品牌独具格调且值得信赖的特质。补充性质的插画传递出的活力和趣味也是品牌形象中十分重要的组成元素。

设计机构：孚曼设计工作室 设计师：乔恩·查普曼·史密斯 摄影：乔恩·查普曼·史密斯 委托方：Mr Toms 咖啡馆

TEL AVIV, ISRAEL

PARTOUT-EVERYWHERE

French Boutique Coffee Shop with Geometric Richness

Partout-Everywhere 咖啡馆 / 以色列，特拉维夫

法式精品与抽象几何的有机结合

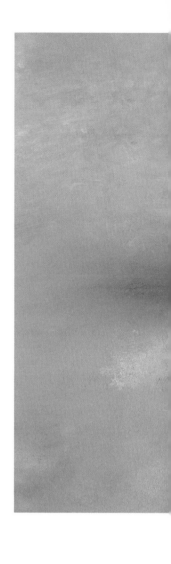

'Partout-everywhere' is a mobile boutique coffee shop, providing the French experience to everyone who pleases. The identity is playing on two levels. The first, the high, French boutique coffee shop, and the second is urbanite seen from above, with all the shades of grey and the geometric richness of it. Both of the levels are expressed throughout the identity, constantly.

Designer: Adi Gaon Photography: Aya Wind Client: Partout Coffee Shop

"Partout-Everywhere" 咖啡馆是一间可移动精品咖啡馆，为顾客提供法式咖啡。品牌形象有两个层面。第一个是高大的法式精品咖啡馆，第二个是不同的灰色和几何形状组成的抽象形象。二者贯穿店铺的整体形象，有机结合。

设计师：阿迪·加翁 摄影：阿亚·温德 委托方：Partout 咖啡馆

JAKARTA, INDONESIA
MONOLOG CAFE

'A Conversation with One Self'

独白咖啡 / 印度尼西亚，雅加达

"与自己进行对话"

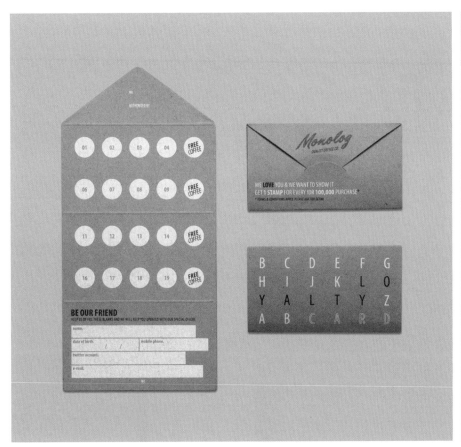
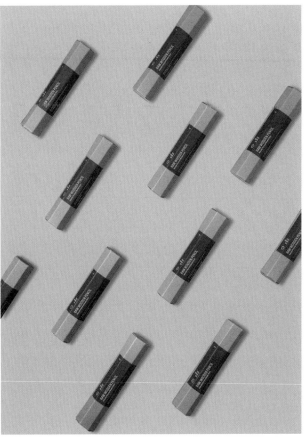

Monolog is a cafe focusing on quality coffee and breakfast menu. The name 'monolog' means 'a conversation with one self'. With this in mind, Brownfox Studio would like to convey the idea of personal ideology and style, not just in terms of food and coffee but also about design and ambience. Therefore the brand itself is very opinionated, and the design team try to translate this into the logo, graphic outlet, and other means of graphic communication.

Design agency: Brownfox Studio Designer: Fergie Tan & Henny Vitri Client: Monolog Coffee

独白咖啡馆供应优质咖啡和早餐。店名中的"独白"意为"与自己进行的对话"。设计师将这一概念以个人意识形态和风格的方式呈现出来,它不只关乎食品和咖啡,还包括设计与氛围。因此设计师试图通过品牌标识、平面设计和其他平面传达手段表达这种观点。

设计机构:Brownfox 工作室 设计师:菲姬·谭,亨尼·维特里
委托方:独白咖啡

AUCKLAND, NEW ZEALAND

ATOMIC

Inspiration Drawn from the Esoteric Tradition of Alchemy

原子咖啡馆 / 新西兰，奥克兰

灵感来自神秘的炼金术

The challenge was to reinvigorate Atomic Coffee's brand identity with a cohesive execution of the brand across all assets. Atomic wanted a premium look, emphasizing the artisan skill and craftsmanship in their growing, roasting and brewing process. The inspiration was drawn from the esoteric tradition of alchemy. The craftsman using secret knowledge and powers to manipulate the elements. The work translates Atomic's passion for uncompromising quality and craftsmanship – while adding a little bit of magic.

Design agency: Fuman Design Studio Designer: Jon Chapman-Smith Photography: Jon Chapman-Smith Client: Atomic Coffee Roasters

本案中设计师面临的挑战是通过具有凝聚力的设计重振原子咖啡的品牌形象。原子咖啡希望得到一流的外观设计,突出种植、烤制和酿造过程中凝聚的工匠技艺。设计灵感来自神秘的炼金术传统。工匠利用秘传的知识与力量控制元素。设计反映出原子咖啡对毫不妥协的品质和工艺的十足热情,在其基础上又增添了一份魔力。

设计机构:字曼设计工作室 设计师:乔恩·查普曼·史密斯 摄影:乔恩·查普曼·史密斯 委托方:原子咖啡馆

TAIWAN, CHINA

COFFEE SMITH

A Storytelling Branding

咖啡匠 / 中国 , 台湾

品牌设计的故事性

Since Smith has the meaning of a craftsman, Coffee Smith can simply be used as Barista. Logo design indicates the quality service provided by professional barista and the meticulous image of a gentleman. Therefore the symbolic bow tie of gentlemen is integrated with coffee mugs to create an intriguing image, expressing the professional spirit of the brand and adding to the humorous nature of storytelling.

Design agency: Vroom Studio   Designer: Celine Shen

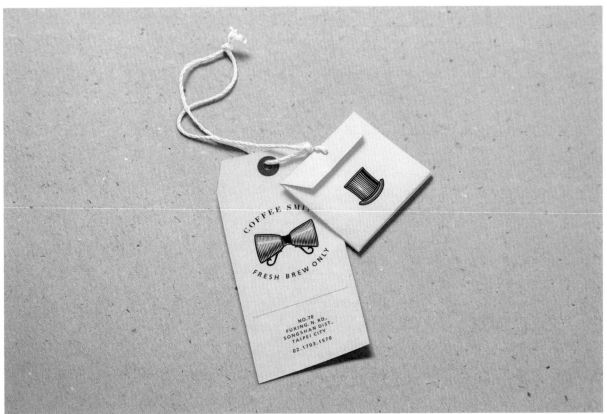

Smith 有工匠的意思，Coffee Smith 则转化为有咖啡师傅的含义。标志设计上希望能表现出专业的咖啡师傅高品质的服务形象，与一丝不苟的绅士印象。因此利用象征着绅士的领结与咖啡杯结合出巧妙有趣的造型，让品牌形象不只透露着专业精神，更带有吸引人的幽默感与丰富的故事性。

设计机构：Vroom 设计工作室 设计师：沈思伶

The brainchild of the man behind Dear Jervois and Little King, Fuman were appointed to create a brand that was both clean and strong yet signified the best of service and excellence in hospitality. The end result reflects the structure and integrity the brand deserves whilst conveying a sense of growth and new beginnings. A rice paddy field subtley denotes the Korean influence found at Major Sprout.

Design agency: Fuman Design Studio Designer: Jon Chapman-Smith Photography: Jon Chapman-Smith Client: Major Sprout

Fuman 受邀为 Dear Jervois and Little King 设计一套简洁有感染力的品牌形象，彰显其卓越的服务。最终的设计成果反映出品牌的结构与整合性，传达一种成长、重新开始的意味。稻场形象体现品牌发展过程中的韩国元素。

设计机构：孚曼设计工作室 设计师：乔恩·查普曼·史密斯 摄影：乔恩·查普曼·史密斯 委托方：Major Sprout 咖啡馆

AUCKLAND, NEW ZEALAND

MAJOR SPROUT

The Appeal of A Wonderful Design

Major Sprout 咖啡馆 / 新西兰，奥克兰

设计赋予的感染力

AMBERG, GERMANY

BA.RO.CO.

Smart and Flexible Logo

ba.ro.co. 咖啡馆 / 德国，安贝格

智能灵活的标识设计

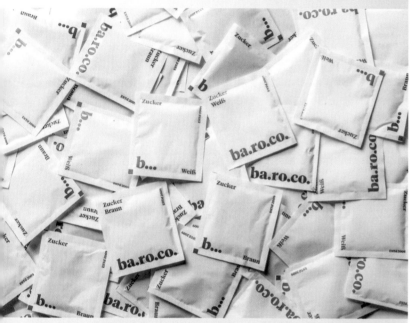

The Bavarian Roasting Company is a small manufacturer of premium coffee. They started with just three coffees because they think it's better to create less, but do it perfectly. They not only offer an exceptional coffee experience to other businesses, but also to customers at their local café, located in the very center of their medieval hometown, where they are known as Café Baroco. Studio Stefanowitsch wanted to capture their philosophy but also play with their catchy nickname. They developed a dual system consisting of a long and a short version of their nickname. The two versions are connected by the same typeface but also, and more importantly, by three dots, one for each of the three coffees they started with. In the long version, the dots abbreviate the full company name, making the origin of the nickname obvious. In the short version, they transform into ellipses, stating: we started small, but there is a lot more to come.

Design agency: Studio Stefanowitsch Designer: Stefan Hoppe Photography: Marcus Rebmann Client: Bavarian Roasting Company

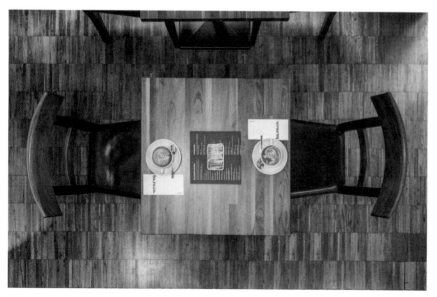

巴伐利亚咖啡烘烤公司是一家生产优质咖啡的小型制造商。公司起步时秉着少而精的理念，只有三种咖啡产品。店内不仅为企业提供绝佳的咖啡产品，其坐落在中世纪小镇中心地带的当地咖啡馆 ba.ro.co. 咖啡也为普通顾客供应高品质咖啡。设计师希望在方案中捕捉品牌精神，也一并在店名的昵称上做些文章。他们为此开发了一个由长短昵称组成的双重系统，两个昵称由相同的字体和三个点联系在一起，三个点代表企业创始之初的三个咖啡品种。长版昵称里，三个点是公司名的缩写，清楚地阐释了昵称的起源。短版昵称里，三个点变成椭圆形，意为：我们从小公司做起，朝着更大更强进步。

设计机构：Stefanowitsch 工作室 设计师：斯特凡·霍普 摄影：马库斯·雷布曼 委托方：巴伐利亚咖啡烘烤公司

WELLINGTON, NEW ZEALAND
COFFEE SUPREME

Design of Irreverent Character

至上咖啡 / 新西兰，惠灵顿

设计里的玩世不恭

Coffee Supreme are one of the largest independent coffee roasters & suppliers in New Zealand, with several of their own cafes including their Auckland HQ. Their rebrand needed to communicate their hand-crafted attention to detail and love of great coffee, as well as what they were already know for; their approachable, supportive, friendly & quirky attitude. The brand uses many hand-drawn illustrations, a utilitarian use of type, a small but strong colour palette & a simplified silhouette version of their logo. The brand now sits well within their existing interior aesthetics, focusing both on their reliability & professionalism, as well as their relaxed, friendly, sometimes even irreverent character.

Design agency: Hardhat Design Creative director: Jenny Miles & Nik Clifford
Designers: Jenny Miles, Nik Clifford, Doug Johns Photography: Paul W Statham, Doug Johns Client: Coffee Supreme

至上咖啡是新西兰最大的独立咖啡烘烤与供应商之一，拥有多处自己的咖啡馆，奥克兰的这间是品牌总店。品牌重塑项目中要求体现手工工艺对细节的关注和对美味咖啡产品倾注的热情，以及品牌已经树立的形象：亲民、友好又略带古怪的品牌精神。品牌形象设计由手绘插图、实用的排版、大胆配色和简洁的标识图案组成。沿用原有的室内美学，兼顾可信性、专业性以及轻松、友好，有时甚至是玩世不恭的特点。

设计机构：安全帽设计工作室 创意总监：珍妮·迈尔斯，尼克·克里夫德 设计师：珍妮·迈尔斯，尼克·克里夫德，道格·约翰斯 摄影：保罗·W·斯坦森，道格·约翰斯 委托方：至上咖啡

LONDON, UK

THE PAVILION CAFE

Café in Greenwich Park

THE PAVILION 咖啡馆 / 英国，伦敦

格林威治公园里的咖啡馆

Identity and branding for a group of cafés and coffee bars in Greenwich Park, London. Each location has a different character and menu, but the whole experience needed to be connected. The design was created to be easily extended and allow materials to be interchangeable.

Design agency: SHO Design Creative director: Jamie McFarlane Designer: Darren James / Jamie McFarlane Client: Creative Taste

本案是为伦敦格林威治公园的咖啡馆进行的品牌形象设计项目。每个地点都有不同的特点和菜单，但整体环境需要联系在一起。设计方案易于扩展延伸，也方便材料之间的互换。

设计机构：SHO设计工作室 创意总监：杰米·麦克法兰 设计师：达伦·詹姆斯，杰米·麦克法兰 委托方：Creative Taste 公司

CAFE AND FASTFOOD

咖啡与简餐店

SYDNEY, AUSTRALIA

LITTLE DELI CO

Warmth like the Wooden Tone

小小食品公司 / 澳大利亚，悉尼

如木质感般的温馨

The Little Deli Co is a boutique deli and café that is all about eating, sharing and loving fine foods, speciality coffee and good conversation. Situated within suburbia Sydney, the prime aspiration is to bring the local community together in a holistic and communal approach.

Design agency: Korolos Design Designer: Korolos Ibrahim
Photography: Shayben Moussa Client: Salam & Hussein Daher

小小食品是一家精品熟食店和咖啡馆，彰显饮食、分享和对精致食物、特产咖啡和促膝长谈的热爱。店铺位于悉尼郊区，追求以公共、全面的方式将当地社区联系在一起。

设计机构：克罗洛斯设计工作室 设计师：克罗洛斯·易卜拉欣 摄影：沙依本·穆萨 委托方：萨拉姆·达希尔，侯赛因·达希尔

GRAZ, AUSTRIA

COFFEE & KITCHEN

A Colourful World in Black and White

咖啡与餐厅 / 奥地利，格拉茨

黑白背景下的色彩世界

Situated in a business district in Graz, Austria the restaurant Coffee & Kitchen brings culinary pleasures to the daily office life. In every detail you can feel that in this project, branding and architecture went 'hand in hand' in order to communicate the following message: Fresh, honest and delicious daily dishes, snacks and coffee for a good price and served in a pleasant ambiance! The colour world in black and white combined with brown determines the interior design as well as the corporate design. The haptics of the brown wrapping paper and brown cardboard (used for menu cards, food packaging...) harmonizes with the furniture made of the natural material, wood. moodley brand identity has consciously avoided the branding printed material however there are different stickers that convey the image of a relaxed and informal restaurant atmosphere (even the signboard symbolizes a sticker). Nice illustrations as well as the mainly used handwriting font, intensify this feeling even more – occasionally interrupted by a reduced straight typography in order to contrast a noble element to the playful one. And that is exactly how presents itself: Relaxed, cool and at the same time, really noble!.

Design agency: moodley brand identity Designer: Nicole Lugitsch Product Photography: Marion Luttenberger Client: coffeeandkitchen Gastronomie GmbH

"咖啡与餐厅"是坐落在奥地利格拉茨商业区的一家餐厅,将美食享受带入日常办公生活。顾客可以在细节设计中感受到品牌和建筑"联手"传达出以下信息:这里提供新鲜、诚实和美味的日常菜肴、零食和咖啡,价格公道,环境怡人。黑白背景下的色彩世界与棕色搭配,奠定了室内设计和企业形象的基调。棕色的包装纸和纸板(用于菜单卡片、食品包装等)的质感与天然木材制成的家具协调统一。品牌形象设计过程中设计师有意避免了使用印刷材料,代替使用休闲餐厅风格的粘贴(甚至招牌都象征粘贴形象)。精美的插画和手写字体强化了这种特点——偶尔穿插直白排版设计,在正式元素与娱乐元素之间形成反差。这体现的正是品牌的核心精神:放松、时尚,同时也非常高贵!

设计机构:moodley品牌形象设计公司 设计师:妮可·卢吉什 摄影:马里昂·鲁腾伯格 委托方:咖啡与餐厅餐饮公司

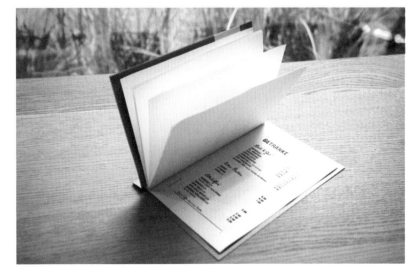

TENERIFE, SPAIN

LA CASITA CAFE

The Colours of Summer

小屋咖啡馆 / 西班牙，特纳利夫

夏日的颜色

La Casita Café is a cozy coffee and restaurant space based in Tenerife, the Canary Islands. The brand promotes freshness, comfort, warmth, & personality. The restaurant desires to make the experience of their customers unforgettable.

Design agency: el estudio Photography: María Laura Benavente Client: La Casita Café

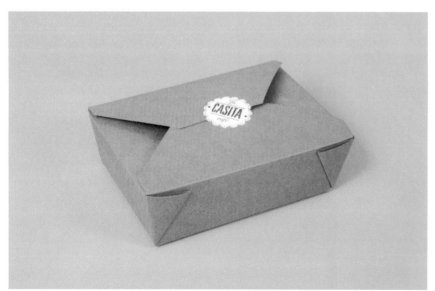

小屋咖啡馆是特纳利夫岛上一家舒适的咖啡馆及餐厅。品牌推崇新鲜、舒适、温馨、个性的理念。餐厅希望为顾客创造难忘的就餐体验。

设计机构：el estudio 工作室　摄影：玛利亚·劳拉·贝纳文特
委托方：小屋咖啡馆

CHATTANOOGA, USA

MILK & HONEY

Illustrated Vitality

牛奶与蜂蜜咖啡馆 / 美国，查特怒加市

插画的活力

Milk & Honey is a locally-owned breakfast, lunch, craft coffee and gelato shop based in Chattanooga, Tennessee. High-quality, homemade and most importantly, real – Milk & Honey was a dream project in every sense of the word. The brand came to life first through the logo and illustrations, moving into interior styling, signage, coffee sleeves, wall menu murals, packaging, labels, gelato carts, apparel and so much more. The goal of the project was to create a space where it's hard to feel anything but happiness when you walk through the door. There is wit and warmth tucked into every nook and cranny. The combination of custom design, copy, illustration and type treatments create a friendly feeling and sense of community. Since the shop launched, they have seen overwhelming success.

Design agency: SeeMeDesign, LLC Designer: Ellen Witt Monen Client: Milk & Honey

牛奶与蜂蜜咖啡馆是田纳西州查特怒加市的一家本地商铺，供应早餐、午餐、手工咖啡和冰淇淋。店铺主打高质量、手工自制，最重要的是货真价实。牛奶与蜂蜜咖啡馆是一项真正洋溢梦想的项目。品牌的活力首先体现在品牌标识和插画设计，随后延续到室内设计、指示牌、咖啡杯隔热套、墙上菜单壁画、产品包装、产品标签、冰淇淋车、工作服等方面。项目的目标是创造一个只能感受到幸福的场所。每个角落都洋溢着智慧和温馨。定制设计、复制元素、插画设计和类型发挥营造出宾至如归的氛围。自开业之日起，咖啡馆就收获了令人惊喜的成功。

设计机构：SeeMeDesign 设计公司　设计师：艾伦·威特·莫能　委托方：牛奶与蜂蜜咖啡馆

One of the most representative icons of Spain, is gastronomy. The north of the country is the biggest cradle of chefs. In a vast scenario of gastronomy proposals, The Bohemian wants to offer a mix of influences for a demanding of new experiences customers. Mixing high quality fast food, a lounge atmosphere and a exclusive cocktail bar.

Design agency: Quim Marin Studio Designer: Quim Marin Client: The Bohemian

美食是西班牙最具代表性的符号之一。西班牙北部走出了无数优秀的厨师。在数量巨大的美食方案中,波西米亚咖啡馆选择供应一组混合菜肴,为顾客提供前所未有的新体验:高品质速食、休闲空间和独享的鸡尾酒吧。

设计机构:奎姆·马林工作室 设计师:奎姆·马林 委托方:波西米亚咖啡馆

IRUN, SPAIN

THE COHEMIAN COFFEE LOUNGE

The Clashing of Blue and Pink

波西米亚咖啡馆 / 西班牙，伊伦

蓝与粉的色彩冲击

LANGENFELD, GERMANY

MAHLWERK

Simple Design and Quality Life

MAHLWERK 咖啡食品店 / 德国，朗根费尔德

简单设计，精致生活

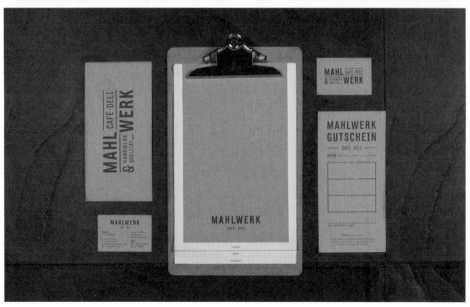

Mahlwerk is a very special Café and Deli based in Langenfeld, Germany. They offer high quality coffee and delicious food, breakfast, salads, sandwiches and cakes – all homemade! Playground were asked to create the corporate design and the input was: simple and honest. A clear and characteristic typography and a minimalistic layout give an impression of what the café represents.

Design agency: Playground Büro für Gestaltung Designer: Anna Pickel & Steffi Zepp Photography: Anna Pickel & Steffi Zepp Client: Mahlwerk Café & Deli

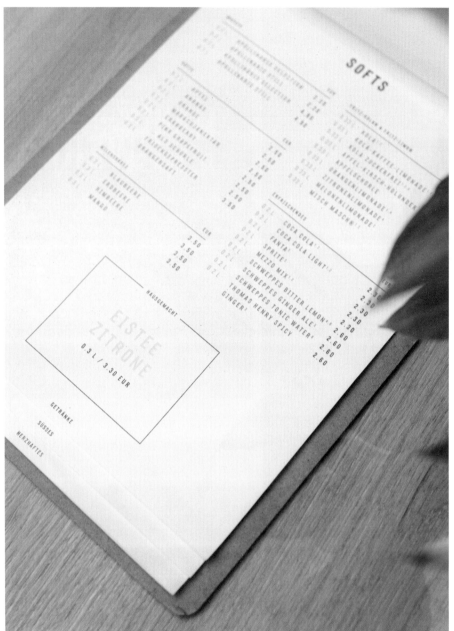

MAHLWERK 是德国朗根费尔德一家非常特别的咖啡馆。店内供应高品质咖啡和美味的食品、早餐、沙拉、三明治和蛋糕，主打全手工制作。设计师受邀为其打造企业形象，遵循简单和诚实的理念。清晰而有特色的字体以及简约的版式设计呈现咖啡馆的整体形象。

设计机构：游乐场设计工作室 设计师：安娜·皮克尔，施特菲·策普 摄影：安娜·皮克尔，施特菲·策普 委托方：Mahlwerk 咖啡食品店

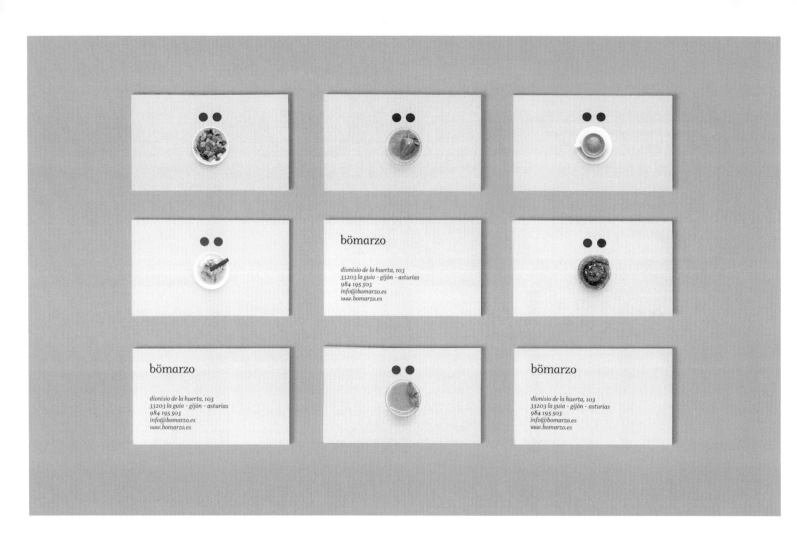

Bömarzo is a new coffee & brunch that it takes its name from the famous Italian gardens (and the famous novel by Mujica Láinez) and whose atmosphere reminds the 'delis' which can be found in countries of northern Europe. The designers create a symbol for identity that synthesizes the entrance of the gardens through the 'o' with dieresis that also is a character of northern European alphabets.

Design agency: atipo® Client: Bömarzo

博马尔佐咖啡馆是一家新开的咖啡早午餐餐厅，店名取自著名的意大利花园（也是穆希卡·莱内兹的著名小说），店内氛围让人联想起北欧国家常见的"熟食店"。品牌形象利用"O"形隔音符号综合了花园入口设计。隔音符号的图案也是北欧使用的字母。

设计机构：atipo 工作室 委托方：博马尔佐咖啡馆

ASTURIAS, SPAIN

BÖMARZO

'o' The Dieresis

博马尔佐咖啡馆 / 西班牙，阿斯图里亚斯

"O"形隔音符号

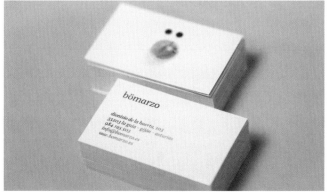

MOSCOW, RUSSIA

CAFE OF JULIA VYSOTSKAYA

Illustrations from Julia Vysotskaya's Books

朱莉娅·维斯托斯卡亚的咖啡馆 / 俄罗斯，莫斯科

朱莉娅·维斯托斯卡亚书中的插画

In Province design studio the designers created a logo and a corporate identity for a new project of the actress and TV presenter, Julia Vysotskaya. In the corporate identity, illustrations from the book by Julia Vysotskaya were used where you can find many recipes of children's food. The light and cozy interior of the café makes you feel at somebody's welcoming kitchen. The place where the pastry chef and the barista are working is one of the main elements in the interior design. Coffee and pastry is made right in the hall.

Design agency: Province Design Studio Designers: Elena Trofimova, Pavel Bogdanov, Evgeny Pakhomov Photography: Alexaner Yarysh Client: The restaurant group of Julia Vysotskaya

本案是设计工作室为女演员兼电视主持人朱莉娅·维斯托斯卡亚的新项目打造的标志和企业形象。设计师在企业形象设计中使用了朱莉娅·维斯托斯卡亚所著书中的插画，里面有很多儿童食品配方。光亮、舒适的室内空间使人感觉似乎置身友人的温馨厨房。糕点师和咖啡师的工作台也是室内设计的一个重要元素，就安排在大厅中。

设计机构：Province 设计工作室 设计师：埃琳娜·特罗菲莫娃，帕维尔·波格丹诺夫，叶夫根尼·帕霍莫夫 摄影：亚历山大·亚雷什 委托方：朱莉娅·维斯托斯卡亚餐饮集团

KYIV, UKRAINE

URBAN HEART

The Simplicity of a Black and White Palette

城市心脏咖啡馆 / 乌克兰，基辅

黑白系的简约态度

Urban Heart is a place of alive communication. There is no wifi, unthinking scrolling and virtual activities. Cafe is open for everyone who knows well what is good coffee, smartmeetings and vivid life of companionship. So, typography strictly built on the rules of cafe.

Design agency: studio 10 Designer: Vallery Shaposhnikoff
Client: International Business Development office 'Strategic'

城市心脏咖啡馆是一个实时沟通场所。店内不提供无线网络,咖啡馆吸引那些真正会欣赏咖啡的人开放,为顾客提供环境优雅的会面空间和放松身心的闲适氛围。排版设计严格按照咖啡馆的品牌标准执行。

设计机构:10号工作室 设计师:瓦勒里·沙伯申科夫
委托方:国际商业发展办公室 "Strategic"

ANTWERP, BELGIUM

TIME OFF

Relaxing and Refreshing Mint Green

"休息时间"咖啡馆 / 比利时，安特卫普

薄荷绿的清新风格

Branding for a new coffee bar in Antwerp (Belgium). Time Off is a cosy place near the city center. A place to relax and enjoy some real good coffee.

Design agency: Babs Raedschelders Client: Time Off
Photography: Babs Raedschelders

本案是为安特卫普一家新咖啡馆进行的品牌形象设计。"休息时间"是市中心一处舒适的休闲场所,是放松心情,享用美味咖啡的好去处。

设计机构:芭布斯·拉德兹彻尔德斯工作室 摄影:芭布斯·拉德兹彻尔德斯 委托方:"休息时间"咖啡馆

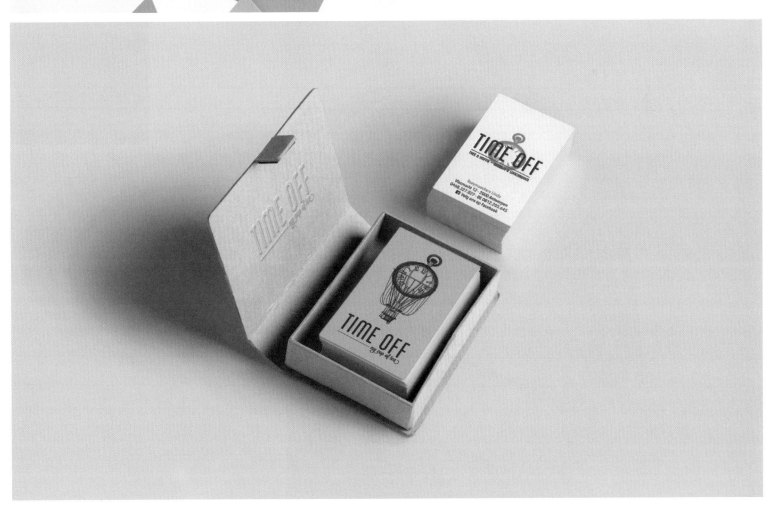

JAKARTA, INDONESIA

PALEM CAFE RESTAURANT

Strong, Unique and Balanced

帕莱姆咖啡餐厅 / 印度尼西亚,雅加达

强大、独特、平衡

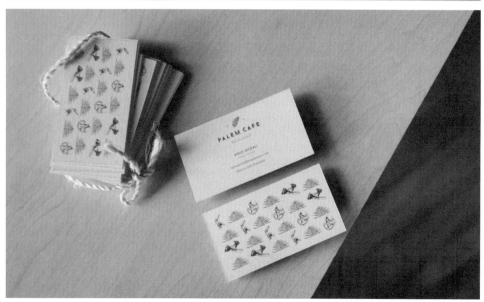

A western restaurant located in Central Park Mall, Jakarta. The selection of the name refers more to the philosophy of palm tree ('Palem' means palm in Indonesian language), which means a tree with a trunk that is strong, robust and proportionate, not faltering when exposed to strong winds. Likewise with that hope, Palem Cafe could be a strong brand that will continue to exist in the competitive world of retail and will become a brand that is able to survive when compared to other brands. The ultimate aim of the redesign of the Palem Cafe restaurant's identity is to make better reflect its identity; strong, unique, balanced, and have good positioning in the eyes of the customers. Another aim is to integrates the concept, identity, and the atmosphere.

Design agency: Elk & Wrakkoon Studio Creative director: Adrian Gozali Designer: Adrian Gozali Photography: Adrian Gozali Client: PT Mitra Adiperkasa, Tbk

本案是位于雅加达中央公园购物中心的一家西餐厅。餐厅名取自与棕榈树有关的哲学理念（"帕莱姆"在印尼语中是棕榈树的意思），意为枝干强壮、匀称，不畏强风的树木。委托人希望帕莱姆咖啡餐厅发展成为同样强大的品牌，能够在竞争激烈的零售世界生存下来。重新设计餐厅品牌形象的最终目的是更好地反映出品牌的特点：强大、独特、平衡，并由此在消费者中树立良好的形象。另外一个目的是实现品牌理念、形象和店内氛围的整合。

设计机构：Elk & Wrakkoon 工作室 创意总监：艾德里安·戈扎利 设计师：艾德里安·戈扎利 摄影：艾德里安·戈扎利 委托方：PT Mitra Adiperkasa, Tbk 公司

BRNO, CZECH REPUBLIC

CAFE CROWBAR

A Strange Angle of Typeface Design

撬棍咖啡馆 / 捷克，布尔诺

字体设计的奇特角度

A Café Crowbar is a small coffee shop located in center of city Brno surrounded by traditional Czech cafeterias. The visual style of the café Crowbar has been inspired by a tool called crowbar. It is a metal bar with a single curved end and flattened points but the rhythm in letters of the word crowbar identifies relaxed atmosphere. The aim of this project was to create a simple logo with easy and cheap application. Reasonably, it was decided to create a brand with simplistic design where no computerized graphic is used.

Design agency: Hany Designer: Hany Zackova
Photography: Richard Procházka and Andrea Navrátilová
Client: Cafe Crowbar

撬棍咖啡馆是布尔诺市中心一家小型咖啡馆,附近有多家捷克传统自助餐厅。撬棍咖啡馆的视觉形象设计灵感正是来自撬棍这种工具。撬棍是一端平坦弯曲的金属条,而这个词本身的韵律透露出店内轻松的气氛。设计的目标是打造一个简单而且应用简单、经济的品牌标识,因而设计方案十分简洁,不涉及电脑图形。

设计机构:阿尼工作室 设计师:阿尼·查科娃 摄影:理查德·普罗查卡,安德里亚·纳芙拉蒂诺娃 委托方:撬棍咖啡馆

GUADALAJARA, MEXICO

LEGACY ROASTERS

Symbolic Use of Graphics

传奇风味咖啡馆 / 墨西哥，瓜达拉哈拉

图形设计的象征印记

Legacy Roasters is an establishment dedicated to the selling and preparation of speciality food and coffee. They know the place where the coffee grew and the different procedures are realized to obtain a variety of flavours, smells and acidity. In the part of the tea Memo & Moi work with one of the best houses of tea in the country named Carabanserai, located in Roma D.F., they provide French tea and realize their own mixtures of excellent quality. Memo & Moi realize the redesign of their identity where they look to keep and stylize the main elements of their old logo, as the top hatted, the gentleman's mustache and the cup of coffee, the result is a clean logo, sophisticated and with an european tendency, to give the classic touch on the composition of the new identity and unique consume experience. On the packaging Memo & Moi use ziploc type bags of rice paper to keep the freshness of the tea and also the coffee, and tag with two stickers, one with the illustration of coffee beans and another one with the picture of a cup of tea.

Design agency: Memo & Moi Brand Consultants Creative director: Guillermo Castellanos & Moisés Guillén Designer: Moisés Guillén & Guillermo Castellanos Client: Carlos Palafox

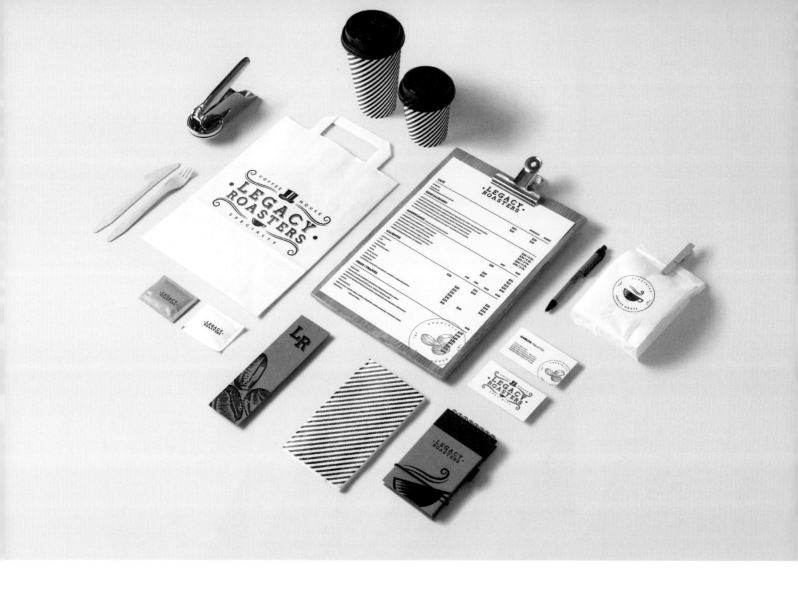

传奇风味咖啡馆是一家专注于制作、经销特色食品与咖啡的食品店。店主对咖啡的产地、工艺和风味了如指掌,并与墨西哥国内最好的茶叶经销商合作,购入高品质的法国茶叶和特制混合茶。设计师受邀对品牌形象进行重新设计,保留原有标识中的绅士胡须和咖啡杯等主要元素,调整设计风格,合成简约、有内涵而略带欧式风格的新标识,为新品牌形象增加经典特质,打造独特的消费体验。包装设计方面,设计师采用了米纸制成的拉链型包装袋,保持茶叶和咖啡的新鲜程度。产品上使用两个贴纸标记,其中一个贴纸上是咖啡豆的插画,另一个是一杯茶的形象。

设计机构:Memo & Moi 品牌咨询公司 创意总监:吉尔勒莫·卡斯特罗,莫伊塞斯·纪廉 设计师:莫伊塞斯·纪廉,吉尔勒莫·卡斯特罗 委托方:卡洛斯·帕拉福克斯

SANTIAGO, CHILE

COLOCHATE CAFETERIA

A Welcoming Sense of Belonging

Colochate 咖啡馆 / 智利,圣地亚哥

温暖的归属感

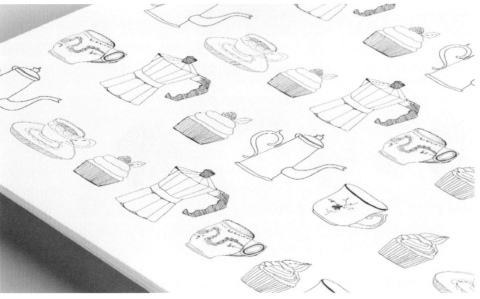

Placed on a business quarter, Colochate Coffee Shop offers delicious meals with a homemade feel for daily lunch. Alima's task was to design the whole identity for Colochate, covering branding, interior space, stationery, advertising and more.

Design Agency: Alima Diseño Creative director: María Vial
Designer: María Vial & Aline Morizon Photography: Aline Morizon Client: Colochate Cafetería

坐落在商务区的Colochate咖啡馆供应有家常亲切感的美味餐食。设计师alima为咖啡馆设计了整体形象，涉及品牌形象、室内设计、文具、广告等。

设计机构：阿利马设计工作室 创意总监：玛利亚·比亚尔 设计师：玛利亚·比亚尔，艾琳·莫利森 摄影：艾琳·莫利森 委托方：Colochate咖啡馆

MONTERREY, MEXICO

CAPULUS TABERNAM

Nordic Impression

Capulus Tabernam 咖啡馆 / 墨西哥，蒙特雷

北欧印象

It means coffee shop in Latin, a place where you can buy many gourmet products, a pack of fresh coffee or take an exotic tea. The aesthetic is based on nordic style.

Design Agency: JRG Creative director: José Roberto González Designer: José Roberto González Client: Capulus Tabernam

店名"Capulus Tabernam"在拉丁语中意为"可以买到许多美食，新鲜咖啡或具有异国风情茶产品的咖啡馆"。设计以北欧风为主。

设计机构 JRG工作室 创意总监：约瑟·罗伯托·冈萨雷斯 设计师：约瑟·罗伯托·冈萨雷斯 委托方：Capulus Tabernam 咖啡馆

LONDON, UK

SMOKE & ROAST

The Colours of Food

Smoke & Roast 咖啡馆 / 英国，伦敦

食物的色彩

Brand identity for new cafe and food offering in London. The identity needed to reflect the slow roasted dishes, with meats cooked over open coals, that form the signature dishes. The colour and typography form the basis of the identity and create an authentic feel of bold flavours.

Design agency: SHO Design Creative director: Jamie McFarlane Designer: Darren James/Jamie McFarlane Client: Creative Taste

本案是为伦敦一家新开的咖啡馆与食品店进行的品牌形象设计。设计需反映店铺用木炭慢火烤制的招牌菜肴形象。配色和字体构成了品牌的基本元素，体现店内食物美味爽口的特点。

设计机构：SHO 设计工作室 创意总监：杰米·麦克法兰 设计师：达伦·詹姆斯，杰米·麦克法兰 委托方：Creative Taste 公司

WALES, UK

PARC PANTRY

Rustic yet Stylish Aesthetic

帕克咖啡食品店 / 英国，威尔士

犀利的现代设计

Parc Pantry is a charming cafe/deli founded on the ideas of local, fresh and bespoke products. The brand relies on the complimentary pairing of custom, hand-crafted type and sharp, modern design to highlight the personality and professionalism of this unique business. The rubber stamped packaging and tags help complete the rustic, yet stylish aesthetic.

Design agency: AT Branding Designers: Andrew Cadywould & Tom Garland Photography: Tom Garland Client: Parc Pantry

Parc Pantry 是一间供应当地新鲜的特色食品的迷人的咖啡餐厅。设计师以互补的搭配,手工制作的优质产品以及犀利的现代设计突出品牌特点和专业度。橡胶加封包装和标签设计使整个项目的时尚度更加完整。

设计机构:AT 品牌工作室 设计师:安德鲁·凯迪伍德,汤姆·加兰 摄影:汤姆·加兰 委托方:帕克咖啡食品店

HANNOVER, GERMANY

KAFFEE KANN ICH.

The Visual Appeal of a Simple Design

Kaffee kann ich. 咖啡馆 / 德国，汉诺威

简洁设计的视觉魅力

Corporate Design for the german brand named 'KAFFEE KANN ICH.' (I can make coffee). Only the the finest locally grown ingredients are used. The choice products are marked and certified with the round brand label.

Design Agency: Cynthia Waeyusoh Kommunikationsdesign
Designer: Cynthia Waeyusoh Photography: Florian Bison
Client: Kaffee kann ich.

本案是为名为"KAFFEE KANNICH（我可以做咖啡）"的德国品牌所作的企业设计项目。该品牌只使用当地出产的最优质食材。推荐产品使用了圆形的品牌标志和认证标识。

设计机构：辛西娅·瓦尔约什设计工作室 设计师：辛西娅·瓦尔约什 摄影：佛罗莱恩·比松 委托方：Kaffee kann ich. 咖啡馆

Hole In The Wall Sandwich Factory is a sandwich bar and café located in the busy legal district of Sydney CBD. A contemporary takeaway and catering joint that is all about awesome sandwiches, good coffee and happy people; and 'awesome sandwiches' means made-to-order, fresh tasting goodness, gently laid between two slices of delicious artisan bread.

Design agency: Korolos Design Designer: Korolos Ibrahim Photography: Korolos Ibrahim Client: Hole in the wall: Sandwich Factory

墙洞三明治工厂是坐落在悉尼繁忙的中央商务区法律区的一家三明治店和咖啡馆。这里提供现代外卖和餐饮服务，人们可以在此找到美味的三明治，优质咖啡和友好的人群。其中三明治是现场制作的新鲜美味，面包均为店内手工制作。

设计机构：克罗洛斯设计工作室 设计师：克罗洛斯·易卜拉欣 摄影：克罗洛斯·易卜拉欣 委托方：墙洞三明治

SYDNEY, AUSTRALIA

SANDWICH FACTORY

The Beauty of Simplicity and Nature

墙洞三明治工厂 / 澳大利亚，悉尼

朴素而自然的美

AUCKLAND, NEW ZEALAND
COMO ST CAFE

A Magical World of Colours

科摩街咖啡馆 / 新西兰，奥克兰

色彩的魔法空间

Brand identity for Como Street Café. Como Street Café is a gourmet cafe based in Takapuna, Auckland.

Design agency: Fuman Design Studio Designer: Jon Chapman-Smith
Photography: Jon Chapman-Smith Client: Como st Cafe

本案是为科摩街咖啡馆进行的品牌形象设计。科摩街咖啡馆是坐落在奥克兰塔卡普纳的一家美食咖啡馆。

设计机构：孚曼设计工作室 设计师：乔恩·查普曼·史密斯
摄影：乔恩·查普曼·史密斯 委托方：科摩街咖啡馆

STAVROPOL, RUSSIA

MOTO MOTO CAFE

Cheerful Madness

Moto Moto 咖啡馆 / 俄罗斯，斯塔夫罗波尔

欢快的疯狂

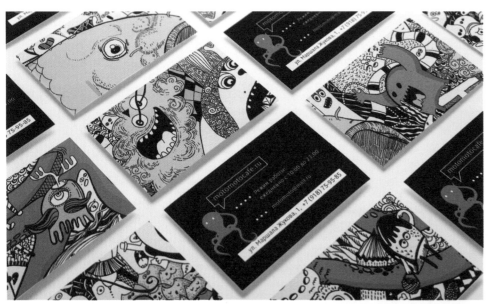

Development of interior elements for cafe of Japanese cuisine. Cafe MOTO MOTO focuses not so much on Japanese cuisine, as on cheerful madness, which is famous for modern Japanese pop culture. Identity places is aspiration to the parody and the sarcastic relation to world around. A non-standard sign in which each letter leads the life. Decorative plates hang above the counter. Signature pattern for countertops are used. Many notable pieces visually convey the basic idea: MOTO MOTO is no place for boredom and dusty templates. Each of the characters is rather mad to correspond to the general style of cafe. The general size of the final image is 7x5 meters. MOTO MOTO is the modern Wonderland.

Design agency: Panfilov&Yushko Creative Group
Designers: Alexej Sobin, Denis Pechersky Photography: Anton Kurashenko Client: Cafe 'MotoMoto'

本案是为经营日式料理的咖啡馆进行的室内装饰元素设计项目。与其说店内特色是日式料理，倒不如说是当代日本流行文化中闻名于世的欢快的疯狂。咖啡馆的设计理念彰显世界的滑稽和讽刺，装饰性盘子悬挂在吧台上方。这里没有无聊和尘土飞扬。每个字符都透露出狂癫，呼应店铺的整体风格。最终的品牌形象尺寸确定为 7×5 米。Moto Moto 咖啡馆是现代版的奇幻仙境。

设计机构：潘菲洛夫与尤什科创意工作室 设计师：阿列克谢·索宾，丹尼斯·派彻斯基 摄影：安东·库拉申科 委托方：Moto Moto 咖啡馆

MILAN, ITALY

MANTRA RAW VEGAN

Taking Away a Seed of Rebirth

曼特拉生素食主义 / 意大利，米兰

"在心中埋下一颗重生的种子"

bīja
(seed)

growth

The architectural, graphic design and communication standards for Mantra came out of the idea of a seed and of essence. Simplicity, then, became the guiding principle for the whole project. It was pursued in the architectural and graphic design, their form and materials. The aim was to deprive everything of unnecessary frills and present rigour, harmony and functionality. 'Anyone who enters is welcome, regardless from its usual diet. Who comes out takes away a smile, an experience of purification, a seed of rebirth.' The environment is young, warm and colourful, elegant but not formal.

Design agency: Supercake Srl Photography: Valerio Gavana

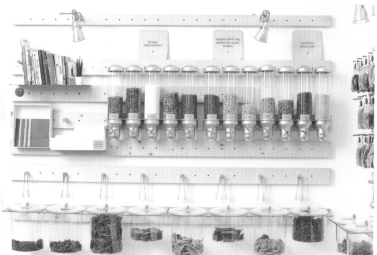

本案中建筑、平面设计的创意来自种子与本质的概念。简约是整个项目的指导原则，贯穿建筑与平面设计，形式与材料。设计师力求去除所有不必要的装饰，获得严谨、和谐、功能的设计。"我们欢迎所有人。人们带着笑容、净化身心的体验离开，在心中埋下一颗重生的种子"。店内环境具有年轻化的特点，温馨又缤纷，优雅但不正式。

设计机构：Supercake Srl 工作室 摄影：瓦莱里奥·加瓦纳

FAENZA, ITALY

CASA E CAFFE MOKADOR

The First Concept Bar of Mokador

Casa e Caffe Mokador 咖啡馆 / 意大利，法恩扎

第一间 Mokador 概念咖啡馆

Mokador is an Italian local roasting born in Faenza in 1967. The concept bar combines the Mokador expertise in coffee and the local food tradition, all under the idea of Casa e caffè (home and coffee), a place where you feel at home and you can taste, learn and buy a high quality coffee. The positioning is for Italian and foreign markets.

Design Agency: 45gradi Creative director: Marina Cattaneo, Silvia Grazioli
Designer: Valentina Ferioli Client: Mokador

Mokador 是 1967 年在法恩扎发明的意大利本土咖啡品种。概念咖啡馆综合了 Mokador 咖啡工艺和当地饮食文化，顾客可以在这里享受到宾至如归的环境，还有机会品尝、学习并购买到高品质咖啡。品牌定位面向意大利本国和海外市场。

设计机构：45gradi 工作室 创意总监：玛丽娜·卡塔内奥，西尔维亚·葛拉索莉 设计师：瓦伦蒂娜·费里索莉 委托方：Mokador 咖啡

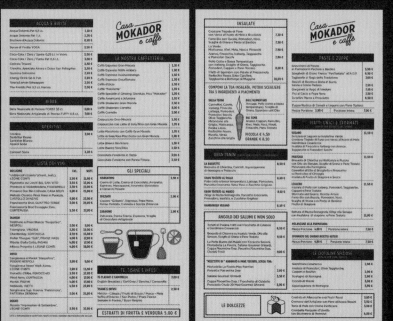

OTHER CREATIVE CAFE

其他创意咖啡馆

BERKSHIRE, UK

CARDINAL CAFE

A Cardinal Bird for Inspiration

北美红雀咖啡馆 / 英国，伯克郡

设计灵感来自北美红雀

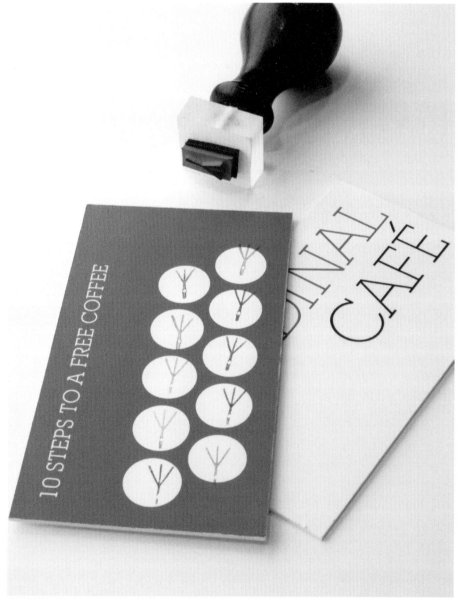

Cardinal Place, a large commercial development in Victoria (500,000 sq ft) opposite Westminster Cathedral, has a vast reception area, within which a cafe was built for the workers on the entire estate. The brief was to create the name and branding for this café, including signage, wall graphics, crockery and other collateral. A character was created using a cardinal bird for inspiration, which not only shares its name with the location but also lent the café its name: Cardinal Cafe. Using a character meant that vibrancy and activity was created in the cafe area, keeping it distinct from the corporate reception area. Graphics were applied using vinyl, which is easily updatable, allowing for the birds to 'move' around the area over time. Applications included: mailers, signage, cups and saucers, napkins and tablecloths.

Design Agency: Hat-Trick Design Designer: Adam Giles Creative Direction: Gareth Howat & Jim Sutherland Client: Land Securities

正对威斯敏斯特大教堂的红衣主教广场有一处宽敞的接待区，这里开设了一家咖啡馆服务整个区域的工作人员。项目要求设计师为咖啡馆设计一个名字以及对应的品牌形象，包括标识招牌、墙壁装饰图案、器具等。设计师以北美红雀为灵感，创造了品牌形象。而北美红雀不仅在读音上与咖啡馆所在的地名"红衣主教"相同（都有cardinal一词），而且咖啡馆也以此命名。设计师在咖啡馆区域使用充满活力的字体，营造与接待区截然不同的氛围。装饰图案由乙烯绘制，便于涂改更新，北美红雀得以在室内"移动"。设计广泛地应用在广告、店铺招牌、咖啡杯、餐巾和桌布。

设计机构：帽子戏法设计工作室 设计师：亚当·贾尔斯 创意总监：加雷思·霍瓦特，吉姆·萨瑟兰 委托方：Land Securities 公司

GUADALAJARA, MEXICO

MISEGRETA

A 'Secret Mix'

MISEGRETA 咖啡馆 / 墨西哥，瓜达拉哈拉

"秘密的混合"

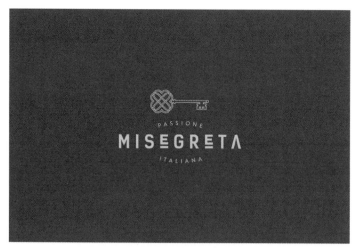

Misegreta is an establishment dedicated to the selling of 100% Italian gelato, desserts, confectionery and specialty coffee. The name is composed of the words in Italian 'miscela' and 'segreto', which in English is 'secret mix', meaning that connects with the logo, a key where on the left side are hearts intertwined which represent the passion, the love and the Italian tradition that are the ingredients of Misegreta's products.

Design agency: Para Todo Hay Fans Designer: Moisés E. Guillén Romero Photography: Moisés Guillén, Guillermo Castellanos

MISEGRETA 咖啡馆专营100%意大利冰淇淋、甜点、糖果和特色咖啡。店名由两个意大利词语"miscela"和"segreto"组成，意为"秘密的混合"。它与品牌的标识设计交相呼应，标识中的钥匙左侧是交织的心形，代表 Misegreta 产品追求的激情、爱和意大利传统。

设计机构：Para Todo Hay Fans 工作室 设计师：莫伊塞斯·E，纪廉·罗梅罗 摄影：莫伊塞斯·纪廉，吉尔勒·莫卡斯特罗

YEREVAN, THE REPUBLIC OF ARMENIA

LOUIS CHARDEN

Illustration Design with Intriguing Stories

Louis Charden 咖啡馆 / 亚美尼亚，埃里温

有故事的插画设计

Backbone were briefed to create a distinct and appealing brand for a new café and bakery, where one can feel the comfort and coziness of French cafes. What comes to your mind when you think of a bakery? Home, family, a fest, childhood or may be a tempting aroma of your mother's cakes? The concept of scent-driven memories led the entire branding process. French origin insinuation was first of all solved through the naming. Backbone developed a brand character - Louis Charden, the founder of the bakery and created a brand story. The identity was based on unique illustrations, unfolding different life events of the brand character– his first love, childish amusements, his first steps in the confectionary industry. No details were overlooked in terms of brand integrity. Today Louis Charden café and bakery are among most favourite sites in Yerevan, where the warm atmosphere naturally makes the visitors a part of brand character life, transferring them into a completely different emotional dimension.

Design agency: Backbone Branding Studio Client: Louis Charden

设计师受邀为这家新咖啡馆和烘焙坊设计一个独特有吸引力的品牌形象，让顾客感受到法式咖啡馆的舒适和惬意。说到烘焙坊时你的第一印象是什么？家、家人、节日、童年，还是母亲制作的蛋糕散发的诱人香气？与香气紧密相关的记忆是整个品牌设计过程的指导概念。店铺命名首先透露出店铺的法国背景。随后设计师创造出有关烘焙坊创始人 Louis Charden 的品牌人物形象以及对应的品牌故事。品牌形象在特色插画的基础上展开，通过生活场景讲述人物的故事——初恋，幼稚的行为，步入糖果行业的始末。设计师尤其注重细节的刻画。如今 Louis Charden 已经是埃里温最受欢迎的咖啡馆之一，这里温馨的氛围让人自然而然地认同品牌精神，体验完全不同的情感。

设计机构：Backbone 品牌设计公司 委托方：Louis Charden

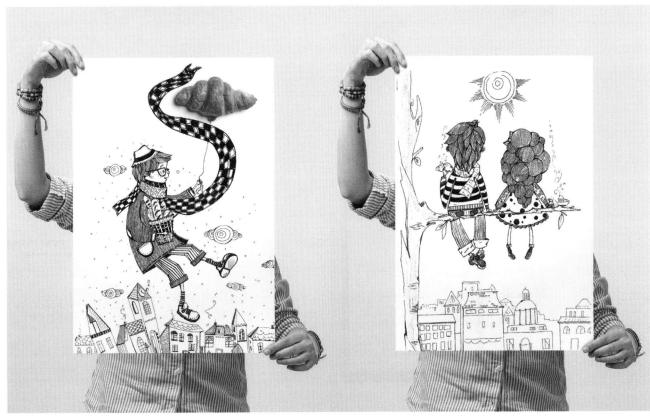

Cloud. is a personal project for coffee shop brand. The designer wanted to create a minimal logo with elements from the sky to express quality within the whole brand and each of its touch points. The idea was to take disparate elements and unify them into whole systems that represent a fresh and playful attitude towards viewers. The outcome is a minimal and modern execution for a fresh and fun brand!

Designer: Wei Xianwen Client: Cloud.Coffee shop

云。是为一个咖啡馆品牌进行的个人项目。设计师希望用天空中的元素创造一个简约标识，表现品牌的核心品质和特点。将不同的元素整合到一起，向消费者呈现出品牌精神与态度。设计成果洋溢着一个新鲜、有趣品牌应有的现代感。

设计师：峗苋汶 委托方：云。咖啡馆

Moodboard

Sofia Pro

ABCDEFGHIJKLMNOPQRSTUVWXYZ
abcdefghijklmnopqrstuvwxyz
123456789

Source Serif Pro

ABCDEFGHIJKLMNOPQRSTUVWXYZ
abcdefghijklmnopqrstuvwxyz
123456789

Font

MILAN, ITALY

CLOUD.COFFEE SHOP

Design with Elements from the Sky

云。咖啡馆 / 意大利，米兰

天空中的元素

WINNIPEG, CANADA

BRONUTS

A Tongue in Cheek Personality

Donuts & Coffee 咖啡馆 / 加拿大，温尼伯

有趣的脸谱

Bronuts' customers include young professionals, corporate offices, and college students. With a bold name, the brand needed to give off a vibe landing in between hipness and approachability. One Plus One engineered a crisp brand identity with a tongue in cheek personality and a simple (but clever) logo, carefully considering each brand extension.

Design Agency: One Plus One Design Designer: Tyler + Jessie Thiessen Photography: Grajewski Fotograph Client: Bronuts Donuts & Coffee

Bronuts 咖啡馆的顾客以年轻专业人士、公司职员和大学生为主。品牌拥有大胆的店名,还需要打造亲民、温馨的氛围。设计师用脸与舌头的形象构建一个清爽的品牌形象。简单而巧妙的标识设计更为考虑到品牌的细节延伸。

设计机构:一加一设计公司 设计师:泰勒·蒂森,杰西·蒂森 摄影:Grajewski 摄影公司 委托方:Bronuts 甜甜圈 & 咖啡馆

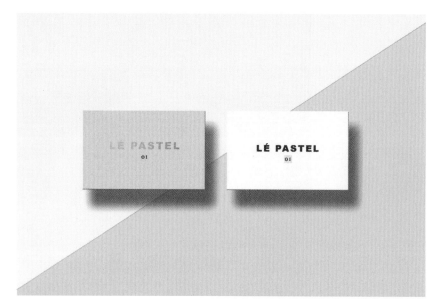

NEW YORK, USA
LE PASTEL

Perfect Balance in a Combination of Colours

彩色粉笔咖啡馆 / 美国，纽约

色彩选择中的完美平衡

Le Pastel is a high-end pastry shop with a variety of delicious desserts. The shop's new identity was designed with one objective, to express quality within the whole brand and each of its touch points. The idea was to take disparate elements and unify them into whole systems that represent a fresh and playful attitude towards customers. The colour plays a big role in this project, since it was meant to create emotion and trigger a sensation of warmth to each and every customer. The combination of neutral colours and pastel colours makes the perfect balance for a clean and sober design. The outcome is a sweet but modern execution for a fresh and fun brand that looks as sweet as it tastes! Bringing happiness in every glance and every bite.

Design agency: Buenas Designer: Claudia Argueta Photography: Claudia Argueta, Diego Castillo Client: Lé Pastel

Le Pastel 咖啡是一家高端糕点店，供应多种美味甜点。全新的品牌形象设计秉承一个目标，即传递品牌的品质和独特之处。设计的基本概念是将不同的元素整合到一个系统中，向消费者呈现一种新鲜又好玩的态度。配色在设计中发挥了重要的作用，起到激发情感，向每位顾客传递温馨氛围的效果。中性色和淡色搭配使用，使简洁、理性的设计达到完美的平衡。设计成果甜蜜可人，又带着作为一个新鲜、有趣品牌应有的现代感。每个细节都洋溢着幸福快乐。

设计机构：Buenas 工作室 设计师：克劳迪娅·阿戈塔 摄影：克劳迪娅·阿戈塔，迭戈·卡斯蒂略 委托方：彩色粉笔咖啡馆

NEW YORK, USA

{t} TELEGRAPHE CAFE

Typography and Symbols

{ t } 电报咖啡馆 / 美国，纽约

字型与符号的发想

The design captures the simplicity and character of Telegraphe Cafe by utilizing only typography and specific symbols to build the entire campaign and to stimulate conversation between the coffee shop and the customer. { t } telegraphe cafe is a student design project. The author and the project have no business, employment or other affiliation with Telegraphe Cafe.

Designer: Chung Hui Pao Photography: Chung Hui Pao

本设计以字型与符号为发想，在品牌识别上描绘出 Telegraphe Cafe 都会简约的风格。整体作品由平面、包装至周边设计也充分运用此视觉语言，串起品牌、空间与人的对话。本设计为学生时期作品，不为该餐厅商业用途使用。

设计师：包忠蕙 摄影：包忠蕙

SEOUL, KOREA

CAFE M

The Symbolic Meaning of the Letter M

M 咖啡 / 韩国，首尔

M 图案的象征意义

This project is for Cafe M, a luxurious and unique cafe with a restful atmosphere in Seoul. Cafe M IS a cafe which provides private spaces and superior quality coffee. Visual identity of cafe M highlights their elegance by using crown shape that stands for royal dignity. The typeface of the brand use bold type to convey a sense of energy and strength.

Design agency: Studio Cervan Designer: Shin Kwangsu
Photography: Shin Kwangsu Client: 5K HOLDINGS FC

本案是为首尔一家兼具豪华和宁静气质的咖啡馆——M咖啡——进行的设计项目。M咖啡为顾客提供私人空间和优质咖啡。咖啡馆的视觉形象设计使用象征皇室尊贵的皇冠图案突出奢华、优雅的特点。设计师使用的大胆字体传达品牌的能量和力量。

设计机构：Cervan工作室 设计师：申光洙 摄影：申光洙 委托方：5K公司

INDEX 索引

& SMITH
024, 088

45gradi
246

+Quespacio
096

Adi Gaon
160

agence 360 degrees
120, 124

Alexia ROUX
110

Alima Diseño
224

ANZI & ONASUP
032

AT Branding
232

atipo®
206

Aysel Sadigova
100

Babs Raedschelders
212

Backbone Branding Studio
090, 256

Bravo
052

Brownfox Studio
162

Buenas
266

Bureau Bumblebee
138

Cast Iron Design
076

Chrome studio
102

Chung Hui Pao
270

Clear
108

Cosa Nostra
072

Cynthia Waeyusoh Kommunikationsdesign
234

Daniele 'Donnie' D'Addario
134

Denis Sharypin
038

el estudio
194

Elk & Wrakkoon Studio
214

ESTABLISHD
150

ESTUDIO INSURGENTE (MÉXICO)
034

Folklore
146

FullFill A rtplication
056, 060

Fuman Design Studio
026, 156, 166, 172, 238

Hany
218

Hardhat Design
180

Hat-Trick Design
250

Isai Araneta
042

Jeong Min Kim
080

Joanna Karwowska, Marek Ejsztet, Piotr Matuszek 020

JRG
228

Korolos Design
084, 188, 236

Koyoox
078

Lacy Kuhn
114

Landini Associates
152

Memo & Moi Brand Consultants
220

MILKnCOOKIES
068

Mind Design
094

Miriam Weiss
130

moodley brand identity
190

One Plus One Design
262

Panfilov&Yushko Creative Group
116, 240

Para Todo Hay Fans
252

PLASMA NODO
064

PLAYGROUND Büro für Gestaltung
202

Pocket
098

Province Design Studio
048, 208

Quim Marin Studio
200

Ramdam Agency
028

Re-public
046

Reynolds and Reyner
142

RG Designs
132

Rice Creative
044

SAVVY Studio
016

SeeMeDesign, LLC
198

SHO Design
184, 230

studio 10
210

Studio Cervan
272

Studio Stefanowitsch
176

Supercake Srl
242

Szani Mészáros
104

Transform Design
126

Vroom Studio
168

Wei Xianwen
260

Wide And Narrow Co., Ltd.
118

图书在版编目（CIP）数据

漫食光：咖啡馆平面与空间设计 /（哥伦）卡洛斯·加西亚，胡书灵编；张晨译 . — 沈阳：辽宁科学技术出版社，2016.9
ISBN 978-7-5381-9819-5

Ⅰ . ①漫… Ⅱ . ①卡… ②胡… ③张… Ⅲ . ①咖啡馆－室内装饰设计 Ⅳ . ① TU247.3

中国版本图书馆 CIP 数据核字（2016）第 117333 号

出版发行：辽宁科学技术出版社
　　　　　（地址：沈阳市和平区十一纬路 25 号　邮编：110003）
印　刷　者：上海利丰雅高印刷有限公司
经　销　者：各地新华书店
幅面尺寸：226mm×240mm
印　　张：23
字　　数：300 千字
出版时间：2016 年 9 月第 1 版
印刷时间：2016 年 9 月第 1 次印刷
责任编辑：杜丙旭　关木子
封面设计：关木子
版式设计：关木子
责任校对：周　文

书　　号：ISBN 978-7-5381-9819-5
定　　价：148.00 元

联系电话：024-23280367
邮购热线：024-23284502
E-mail：1207014086@qq.com
http://www.lnkj.com.cn